Spatial Data Modelling for 3D GIS

Alias Abdul-Rahman · Morakot Pilouk

Spatial Data Modelling for 3D GIS

Springer

Dr. Alias Abdul-Rahman
Department of Geoinformatics
Faculty of Geoinformation
Science and Engineering
Skudai 81310
Johor
Malaysia
alias@fksg.utm.my

Dr. Morakot Pilouk
ESRI
380 New York Street
Redlands 92373-8100
USA
mpilouk@esri.com

Library of Congress Control Number: 2007932286

ISBN 978-3-540-74166-4 Springer Berlin Heidelberg New York

Springer is a part of Springer Science+Business Media
springer.com

Typesetting: by the authors and Integra, India
Cover design: deblik, Berlin

Printed on acid-free paper SPIN: 12038497 5 4 3 2 1

Preface

This book is based on research works done by the authors at the University of Glasgow, Scotland, United Kingdom and the International Institute for GeoInformation Science and Earth Observation (ITC), The Netherlands in 2000 and 1996 respectively. We were motivated to write the book when we began a joint research work in 1992 for our postgraduate theses on Digital Terrain Modelling (DTM) data structuring and eventually DTM software development based on triangular irregular network (TIN) data structure. We realized then that many aspects needed to be addressed especially if an advanced geo information system (GIS) such as 3D GIS system was to be realized. Research in 3D GIS is getting growing in interest and this has really motivated us to do more experiments in the 3D domain. One of the most current interesting issues is spatial data modelling for 3D GIS.

We would like to thank our former supervisors, Dr Jane Drummond of University of Glasgow and Dr Klaus Tempfli of ITC. Various helps received from friends and colleagues at both institutions are also acknowledged. Special thanks go to Mohamad Hasif Nasaruddin, a postgraduate student at the Dept of Geoinformatics, Faculty of Geoinformation Science and Engineering, Universiti Teknologi Malaysia (UTM), Johor, Malaysia for his patient in formatting the manuscript.

This book aims to introduce a framework for spatial data modelling for 3D GIS and it is specifically written for GIS postgraduate level courses. Postgraduate students, researchers, and professionals in Geo Information (GI) science community may find this book useful and it may provide some insights in various spatial data modeling problems. We hope that this book will serve as one of the useful resources in 3D GIS or 3D geoinformation research.

Alias Abdul-Rahman (UTM, Johor, Malaysia)
Morakot Pilouk (ESRI, Redlands, CA, USA)

2007

Contents

Chapter 1 INTRODUCTION

1.1 Why does 3D GIS Matter?

Next generation of Geo Information System (GIS) requires a new way of spatial data modelling. We call the next generation of GIS 3D GIS. Fundamentally, a new digital model has to be developed or established. Exploiting digital computing technology to improve the quality of life, or to prevent or mitigate hazards or disasters, would first require the construction of a model in digital form of the part of the earth and its environment. Such a model, a simplified description of complex reality, can conveniently be used, stored, managed, maintained, distributed, and transported. Even a complex model may be stored on a small scale, in diskettes, tape cartridge or CD ROM, or transmitted via communication networks. A digital model contains spatial and non spatial aspects of reality and provides a basis for operation and communication among the interested parties. A model distinguishes objects an object, or a set of objects, comprises the elements of reality under investigation. Spatial aspects are those related to shape, size and location that pertain to geometric properties. Non spatial aspects include name, colour, function, price, ownership, and so forth, often referred to as thematic properties. Spatial aspects of reality can be well and economically represented in the form of graphics, whereas non spatial aspects, in many cases, can better be represented in text. Graphic representation facilitates rapid understanding of the situation in reality, permitting high level abstraction or description about neighbouring relationships, while the textual representation is more suitable for aspects that cannot be graphically described. A digital model must be capable of relating these two representations. Creating such a model as an artificial construction of reality in a computing environment requires a tool set exploiting the technology both of computer graphics (CG) (Sutherland, 1963, 1970; Foley et al., 1992; Watt, 1993) and database management (DBMS). Geographic information systems (Burrough, 1986; Maguire et al., 1991), and computer aided design (CAD) are examples of such tools. The essential difference between GIS and CAD is the handling of the spatial aspects rather than the non spatial aspects.

Geographical Information Systems (GISs) represent a powerful tool for capturing, storing, manipulating, and analysing geographic data. This tool is being used by various geo-related professionals, such as surveyors, cartographers, photogrammetrists, civil engineers, physical planners (urban and rural), rural and urban developers, geologists, etc. They use the tool for analysing, interpreting, and representing the real world and understanding the behaviour of the spatial phenomena under their respective jurisdictions. Almost all of the systems used by the geoinformation community to date are based on two-dimensional (2D) or two-and a half-dimensional (2.5D) spatial data. In other words, one may find difficulty processing and manipulating spatial data of greater dimension than 2 in the existing systems, resulting in inaccurate or at least very incomplete information. Furthermore, manipulating and representing real world objects in 2D GIS with relational databases are no longer adequate because new applications demand and increasingly deal with more complex hierarchical spatial data than previously supported by the relational model. It has been suggested in the literature that the abstraction of complex spatial data could be handled more effectively in object-oriented rather than in relational database environment (Egenhofer and Frank, 1989; Worboys, 1995).

The limitations of the current 2D GISs, especially in geoscience, have been reported in the literature by Jones (1989), Raper and Kelk (1991), Rongxing Li (1994), Houlding (1994), Bonham-Carter (1996), and Wei Guo (1996). The limitations mentioned relate to data dimensionality and data structures. Single valued z-coordinate data such as a point (x, y coordinates) with the z-coordinate representing height presents no data handling difficulty in such systems, but it is inadequate for data with multiple z-values (Bonham-Carter, 1996; Raper and Kelk, 1991) such as ore bodies and other important three-dimensional real world entities. A major impediment to establishing 3D GISs is associated with inappropriate spatial data structures, as reported in Jones (1989) and Rongxing Li (1994). These two authors have proposed voxel data structures for 3D data as a solution to the data structuring problem, but no real operational system was developed based on the structure. The problem was also highlighted in the geological field by Houlding (1994). True representations and spatial information, for example sub-surface 3D objects, could not be successfully achieved with 2D systems. 3D visualisation tools alone (for example Advanced Visualization System (AVS), Voxel Analyst of Intergraph, and other Digital Terrain Model (DTM) packages) were not able to truly manage such data as demanded. For example Wei Guo (1996) experimented with the 3D modelling of buildings by using Molenaar's (1992) formal data structure in the relational database environment together with AutoCad as a 3D visualization tool; AutoCad was used to generate the building models. In the literature, a common suggestion has been that the existing GISs were able to handle most of the 2D spatial data, but had difficulty in handling 3D spatial data and beyond, therefore, an extension of the existing systems to at least a third-dimension (3D) is one of the solutions suggested by GIS researchers.

Another observation is that the literature cites no work on three-dimensional GIS coupled with object-oriented technology. Given that the weakness of conventional off-the-shelf 2D or 2.5D GISs are revealed when three-dimensional real world entities are considered, it is understood that object-orientation and three-dimensionality are not more often jointly considered. Some works have focussed on 3D issues such as work reported in Fritsch and Schmidt, 1995; Kraus, 1995; and Fritsch, 1996. But all of these attempts were based on the relational database environment. Therefore, this research monograph looks at both 2D and 3D spatial data modelling and the development of a geoinformation system using relational and object-oriented technology to attempt to solve 3D problems in the GIS environment.

1.2 The Need for 3D GIS

We live in a three dimensional (3D) world. Earth scientists and engineers have long sought graphic expressions of their understanding about 3D spatial aspects of reality in the form of sketches and drawings. Graphical descriptions of 3D reality are not new. Drawings in perspective view date from the Renaissance period (Devlin, 1994). 3D descriptions of reality in perspective view change with the viewing position, so their creation is quite tedious. Traditional maps overcome this problem by using orthogonal projections of the earth. However, they offer a very limited 3D impression.

These traditional drawings and maps reduce the spatial description of 3D objects to 2D. Using computing technology, however, knowledge about reality can be directly transferred into a 3D digital model by a process known as 3D modelling. A 3D description of reality is independent of the viewing position. Adequate cover of the aspects of reality under investigation requires its understanding from many different viewpoints. The disciplines of geology (Carlson, 1987; Bak and Mill, 1989; Jones, 1989; Youngman, 1989; Raper and Kelk, 1991), hydrology (Turner, 1989), civil engineering (Petrie and Kennie, 1990), environmental engineering (Smith and Paradis, 1989), landscape architecture (Batten, 1989), archeology, meteorology (Slingerland and Keen, 1990), mineral exploration (Sides 1992), 3D urban mapping (Shibasaki et al., 1990; Shibasaki and Shaobo, 1992), all draw on 3D modelling for the efficient completion of their tasks.

A 3D model is the basis of a system providing the functionality to accomplish the task in hand. Scott (1994) has summarized the work of Bak and Mill (1989), Fisher (1993), Kavouras and Masry (1987), Raper (1989), Raper and Kelk (1991), and Turner (1989), to provide a set of functions that can be expected from 3D modelling. These studies should provide the means for constructing a 3D model from disparate inputs; permit the maintenance of existing models; facilitate effective 3D visualization with, for example, orthographic, perspective or stereo display with hidden line/surface removal, surface illumination, texture mapping; spatial analyses enabling the calculation of volume, surface area, centre of mass, optimal path as well as spatial and non spatial search and inquiry.

CAD is a typical CG tool for 3D modelling used in car, machinery, aircraft and spacecraft designs, the construction industry, and architecture. CAD focuses on the geometric aspect of the model and its 3D visualization. An example would be a perspective view with hidden line and surface removal, surface illumination, ray tracing, and texture mapping. The question arises whether CAD can support all the tasks required in the disciplines listed above. Attempts have been made to use CAD for tasks in earth sciences requiring 3D modelling and functionality. However, it cannot immediately be assumed that CAD is suited to these tasks, for the following reasons.

- CAD was developed to solve problems in the design of man made objects with well or predefined shapes, sizes, spatial relationships and thematic properties. CAD does not provide the tools for data structuring, or dealing with objects lacking such well-defined shapes, sizes, spatial relationships and thematic properties. Neither is it capable of analysing spatial relationships, nor coping with the disparate data sets and uncertainty typically encountered in GIS. For example, CAD will not reliably maintain the neighbourhood relationships between objects important in earth science analyses, because these relationships may not be considered significant in the design.
- Designing an object, such as a building, is a subjective matter. All aspects of objects and their relationships have to be decided by a human designer; there is little that can be automated. Earth science applications seek to model existing objects, with shapes, sizes and interrelationships outside human control. Here, automation is desirable because of the large number of objects involved. Some relationships important for spatial analysis have to be created automatically. CAD does not usually provide a function for this kind of automation.

- CAD starts the object definition from 3D. When objects are broken down in 2D components, the relationships between them are known. Earth science applications typically model components of reality separately, mostly in 2D, and are dominated by the application view, available tools and information. The components have to be combined and their interrelationships discovered at a later stage. This is quite difficult, since CAD does not usually provide sufficient tools to derive the relationships between the separate components.
- CAD creates a complex object by combining several components possessing such simple geometry as a cube, cylinder, or sphere. The operations of transformation, union, and intersection can be readily applied to such components to obtain the complex object. Earth science applications usually treat a complex object as a whole. Decomposition into primitives is comparable to reverse engineering, the opposite of CAD. The modelling approach used by CAD may not therefore always be suitable for earth science applications. Geometric primitives of an even lower level, such as points and lines, are needed to represent complex reality beyond man made objects.

These geometric primitives also determine the related operations which CAD may not be capable of providing.

A more suitable tool for earth science applications would be a GIS providing a 3D modelling capability, that is to say, a 3D GIS. At the time of writing, a GIS capable of providing the functions listed above list with full 3D modelling capability is not commercially available. Most GISs still limit their geometric modelling capability to 2D so that the 3D representation, analysis and visualization provided by CAD are not possible. Most endeavours to model the third dimension can be found in the representation of terrain relief and in digital terrain models (DTM). DTM can facilitate spatial analyses related to relief, including slope, aspect, height zone, visibility, cut and fill volume, and surface area, and the 3D visualization of a surface, as in a perspective view. However, the basis of DTM is a continuous surface with a single height value for every planimetric

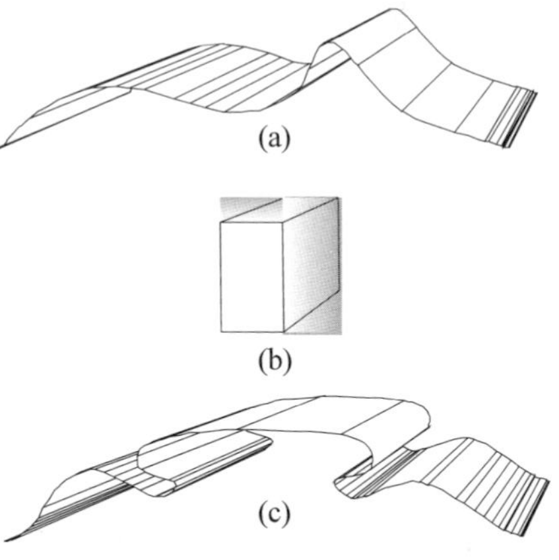

Fig. 1.1 Single-valued surface (a), 3D solid object (b) and multi-valued surface (c).

location (see Figure 1.1a). DTM cannot accommodate a 3D (solid) object, or a surface with multiple height values at a given planimetric location (see Figure 1.1b and Figure 1.1c, respectively).

Although raster-based systems which could be regarded as 3D GISs are available, they may not be able to maintain the knowledge about reality available in the original data set. This knowledge may be lost because of problems in resolution and resampling. As a remedy, the original data set would have to be stored separately from the model, for example, for:

- recreating the model if the result proves to be unsatisfactory because of unsuitable mathematical definition
- creating another model with different resolution
- merging with another data set to create a new model
- archiving as a reference to, or evidence of, the model.

These activities imply the need to store original data in an appropriate structure ready for future use. Necessary information about the data should be attached to each data element. In DTM for instance, information that a line is a breakline should be kept because it will have an impact on the interpolation. Similarly, other information can be attached which influences data handling strategies.

Since neither CAD nor GISs can at present fulfil the requirements of earth science applications, further research and development of a 3D GIS would seem appropriate.

Who needs 3D GIS?

As in the popular 2D GIS for 2D spatial data, 3D GIS is for managing 3D spatial data. Raper and Kelk (1991), Rongxing Li (1994), Förstner (1995), and Bonham-Carter (1996) present some of the three dimensional application areas in GIS, including:

- ecological studies
- environmental monitoring
- geological analysis
- civil engineering
- mining exploration
- oceanography
- architecture
- automatic vehicle navigation
- archaeology

- 3D urban mapping
- landscape planning
- defence and intelligence
- command and control

The above applications may produce much more useful information if they were handled in a 3D spatial system, but 3D spatial objects on the surface and subsurface appear to demand more complex solutions (e.g. in terms of modelling, analysis, and visualization) than the existing systems can offer.

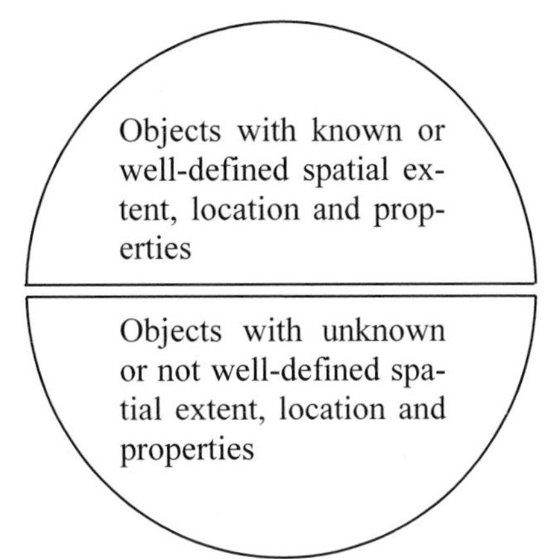

Fig. 1.2 Two types of real world objects with respect to their spatial extent.

1.3 The Need for 3D Spatial Data Modelling

In addition to the problem of creating a system capable of offering 3D modelling and functionality, there is a further problem concerning the type of 3D model chosen as the basis for 3D GIS. The model contains knowledge about reality, so we consider below the types of real world objects it must represent. Two kinds of real world objects may be differentiated in terms of prior knowledge about their shapes and location, as shown in Figure 1.2. In reality, objects from the two categories coexist. Traditional GIS models the objects of each category independently with the result that two separate kinds of systems or subsystems have been developed.

Raper (1989) has also defined these two categories of objects. The first category, regarded as 'sampling limited', is for objects having discrete properties and readily determined boundaries, such as buildings, roads, bridges, land parcels, fault blocks, perched aquifers. The second category, known as 'definition limited', is for objects having various properties that can be defined by means of classification, using property ranges. For example, soil strata may be classified by grain-size distribution; moisture content, colloid or pollutant in the water by percentage ranges; carbon monoxide in the air by concentration ranges, and so forth. Molenaar (1994a) regards these objects as 'fuzzy spatial objects'.

Separate modelling of these two categories of objects tends to contradict the reality, which leads to difficulties in representing their relationships. Such a question as, 'how many of the people working in a 50-storey office building are affected by polluted air generated by vehicles in nearby streets during rush hours?' cannot be answered until the two separate models are combined, as shown in Figure 1.3. Modelling them together with more accurate representation of their relationships in the 3D environment requires the integrated 3D modelling.

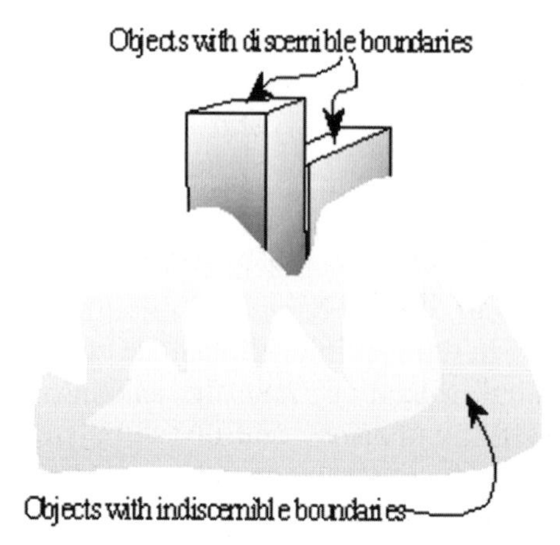

Fig. 1.3 An example of two types of real world objects

Note also that the properties of an object may be well defined in some specific dimensions and ill defined in others. For example, given a DTM data set representing a surface, the planimetric extent of regions at the elevation of 100 metres above mean sea level cannot be defined until the result of interpolation based on a mathematical definition (for example, linear interpolation) is obtained. That is to say, although the spatial extent of this region may be known in the z-dimension, the spatial extent in planimetry (x, y) has still to be discovered. The model must contain the aspect allowing the appropriate operation, such as interpolation or classification, if the required description of the properties of an object is to be obtained.

Apart from the problem of the separate modelling of the two types of objects, there remains the further problem of the separate modelling of an object's components. These components are relief and planar geometry associated with thematic properties. This separation has resulted in independent systems and data structures, DTM and 2D GIS, respectively. The consequences are data redundancy, which may lead to uncertainty when the two data sets are combined and only one data set has been updated.

DTM can facilitate several GIS analyses and visualization taking into accounts the third dimension. The spatial information stored in DTM and in GIS, however, can only be related through coordinates. This implies that relationships between different components may not be properly represented because of metric computation instead of topology. To overcome this, information derived from DTM must be converted into a form GIS can recognize. For example, information about a slope or height zone must first be converted into a thematic layer of GIS for further overlaying before the spatial analysis can be carried out. Imagine having information about the relief, planimetry and themes integrated into one model, so that conversion of such information as slope, height zone and so forth were no longer necessary. Such a question as, 'which land parcels are subject to one-metre flooding?' could be answered from one model. Integrated modelling of this kind is evidently also required for 3D GIS.

1.4 Problems Associated with Spatial Modelling for 3D GIS

Establishing a 3D GIS while taking into account the integration of the necessary components and different types of objects requires the solution of the following problems related to the spatial model representing reality:

1) Design of a spatial model
   - design of an integrated data model, or a scheme, permitting the derivation of a unified data structure capable of maintaining all the components of the geometric representation of real world objects, whether obtained from direct measurements or from derivations, in the same database. Each geometric component must be capable of representing a real world object differently understood by different people.
2) Construction of a spatial model
   - development of appropriate means and methods for 3D data acquisition;
   - coordinate transformation into common georeferencing when different components are to be included into one database;
   - development of a data structuring method that unites the data from various inputs of multi sources into an integrated database capable of being maintained by a single database management system;
   - design of thematic classes to organize representation of real world objects with common aspects into the same category;
   - solving the uncertainty arising from discrepancies from different data sets during the integration process and converting the uncertainty into a 'data quality' statement to be conveyed to the end user.
3) Utilization of a spatial model
   - utilization of existing components, such as 2D data and DTM (backward compatibility) and preparation of those components for future incorporation into the higher-dimension model (forward compatibility) to save the costs of repeating data acquisition.

- development of additional spatial operators and spatial analysis functions;
- development of maneuverable graphic visualization permitting the selection of appropriate viewpoints and representation enabling convenient, adequate uncovering of the details of objects stored in the database;
- design of 3D cartographic presentation of information, including name placement, symbol, generalization, etc.;
- design of a user interface and query language allowing users access to the integrated database;
- development of a spatial indexing structure that speeds up data retrieval and storage processes for the integrated database, including specific (database) views for each user group and guidelines keeping these views updated according to the core database;
- development of tools for navigating among different models stored in databases at different sites and computing platforms.

4) Maintenance of spatial model
   - design of updating procedures, including the development of consistency rules ensuring the logical consistency and integrity of the integrated database, especially during the updating process.

1.5 Previous Work

The status and progress of research in the 3D GIS field within the scope of this monograph and the identification of solutions and remaining problems are made clear from the following review of previous work.

The development of data models for a 3D GIS has branched in two directions. The first is the full 3D approach that looks directly into the design of a data model suitable for 3D GIS. Molenaar (1989) proposes a formal data structure (FDS) for a 3D vector map which may be regarded as a generalization of the 2D version of FDS. Shibasaki and Shaobo (1992), Rikkers et al. (1993), Bric (1993), Bric et al., (1994), and Wang (1994) have reported experimental use of 3D FDS.

The second approach comes from the viewpoint referred to as the 'integration of DTM and GIS'. DTM became a discipline in its own right in the late 1950s (Miller and Laflamme, 1958). Fritsch (1990) has recognized the work of Makarovic (1977) as a proposer of this integration. Males (1978) though not addressing the integration issue, demonstrated the use of a triangulated irregular network (TIN) permitting the attachment of thematic information with elements of TIN in the ADAPT system.

Further steps towards this integration date from the late 1980s, when DTM became an essential part of many complex spatial analyses in GIS in erosion and slope protection, flood protection, the planning of irrigation for agriculture, the geometric correction of remotely sensed images, and so forth. Würländer (1988) investigates some strategies for integrating DTM into GIS. Sandgaard (1988) describes an attempt at integrating DTM into the Dangraf system to facilitate the production of maps with contour lines. Mark and colleagues (1989) report an approach to interfacing a GIS based on quadtree (Samet 1990) with a regular grid DTM for display or analysis. Ebner and colleagues (1990) propose the 'subroutine interface' which was implemented in the program package HIFI-88. Subroutines for interactive editing of GIS are provided for updating DTM, for example, point insertion and deletion, and the change of coordinates in planimetry and height while databases of DTM and GIS remain separate. Ebner and Eder (1992) report drawing on this approach to the facilitation of spatial analysis, using the HIFI-GIS interface with the SICAD-Hygris System to analyse forest damage in terms of such relief parameters as height, slope and exposition. Fritsch (1990) reports the realization of integration at the data structure level. Rather than a full 3D data structure, he suggests an approach that separates two geometric databases for terrain and situation data from another for thematic data. These three data sets are managed within one object oriented database environment. Fritsch and Pfannenstein (1992a) weigh the advantages and disadvantages of integration based on regular-grid, TIN and a hybrid of both. Fritsch and Pfannenstein (1992b) extend this comparison to the layer (organizing different themes in specific layers) and object class (organizes objects into a hierarchy) approach.

An issue in spatial modelling concerns the representation of spatial relationships. Egenhofer (1989), Jackson (1989), Kainz (1989), and Pigot (1991) have described the representation of spatial relationships between objects in 2D and 3D space, based on sound mathematical concepts.

Regarding the issue of model construction, CAD and most CG software packages provide interactive tools for the manual construction of models of objects with discernible boundaries. Manual construction is labourious and the method would not cope with large numbers of objects. For objects with indiscernible boundaries, significant progress has been made in computational geometry based on 2D and 3D Voronoi tessellation (Voronoi 1908, Thiessen 1911, Dirichlet 1850), in the construction of TINs, and tetrahedral networks (TEN). Watson (1981), Avis and Bhattacharya (1983),

Edelbrunnner and colleagues (1986), Tsai and Vonderohe (1991), Midtbø (1993) have all suggested methods for the construction of TEN based on Delaunay triangulation criteria (Delaunay 1934). These methods were extensively applied long ago to the construction of TIN (Shamos and Hoey 1975, Lawson 1977, Lewis and Robinson 1978, Sibson 1978, McCullagh and Ross 1980, Lee and Schachter 1980, Bowyer 1981, Watson 1981, Mirante and Weingarten 1982, Maus 1984, Dwyer 1987, Sloan 1987, Macedonio and Pareschi 1991, etc.). However, these developments are quite independent of GIS.

For the issue of the exploitation of the 3D model, considerable progress has been reported in two other disciplines exploiting CG technology, namely CAD and virtual reality (VR). CAD and VR provide a realistic visualization capability, that is to say, perspective display with hidden line and surface removal, shading and surface illumination, ray tracing, and texture mapping. In addition, VR provides high interactivity within the concept of 'functional realism', allowing the user to manipulate and interact with virtual objects stored in the computer's database as in reality. For instance, the user can 'grab' a virtual object displayed on the computer screen, using the interfacing device called a 'data glove' which sends feedback to the user's hand (for example, a pulse, or vibration) as soon as the virtual object is virtually touched. Developments in this direction are also quite independent of GIS.

The status of the research in 3D GIS and the most relevant remaining problems can be summarized in the following statements:

- The full 3D approach, 3D FDS, does not support well the modelling of real world objects whose boundaries cannot be directly determined. Further extension to cover this issue is therefore needed.
- Progress made by the integration approach can only achieve solutions for surface related objects with little support from theoretical concept of spatial modelling. Extension of this approach to full 3D based on sound spatial mathematics is required.
- Efficient methods for data acquisition, data structuring, database creation and updating with respect to 3D GIS have yet to be developed.
- The incorporation into 3D GIS of independent developments in 3D visualization and 3D geometric construction, whether manual (interactive 3D graphical editing) or automatic (3D Voronoi and tetrahedral network), needs further research.

1.6 Background to the 3D GIS Problem

In geomatics or geoinformatics we consider real world objects exist in three-dimensional (3D), thus it is desirable to have a system which is able to store, handle, manipulate, and analyse objects in a 3D environment. As mentioned in the previous section, the current popular GIS software handles, manipulates, and analyses geographic data in 2D or 2.5D, thus using this system to manipulate 3D data full (particularly multiple Z coordinates) information about real world objects may not be appropriate. Therefore, the 2D GIS (or 2.5D GIS) needs to be extended, i.e. to 3D GIS. Only within the last decade has 3D GIS begun to be discussed in the GIS research community (Raper and Kelk, 1991; Rongxing Li, 1994). The development of this particular GIS approach seems to be relatively slow due to the lack of proper spatial data models and data structures, and the lack of a comprehensive theory of object relationships and data basing for the 3D environment (Wei Guo, 1996). Attempts have been made to develop 3D GIS by Li *et al.* (1996), Pilouk (1996) and Qingquan Li and Deren Li (1996). Li's use an octree approach for 3D subsurface geological modelling, Pilouk uses a 3D TIN approach for regular features on the terrain, while a combination of octree/tetrahedron was proposed by Qingquan Li and Deren Li. Others have used Constructive Solid Geometry (CSG) and Boundary-representation (B-rep) approaches (Cambray, 1993; Cambray and Yeh, 1994; Bric, 1993; Bric et al, 1994; and Zeitouni et al, 1995). All of this work were based on regular shaped objects, which were man-made, and relational data basing. Nonetheless, there appears very little published work on the modelling of 3D objects including natural objects, e.g. forests, plants, water bodies, and other natural subsurface features using the object-oriented (OO) approach. Recent research (Rongxing Li, 1994 and more recently Fritsch, 1996) in this domain have suggested that 3D spatial data modelling, structuring and data basing with object-orientation leads to better 3D GIS. This suggestion seems mainly arise from the complexity of 3D spatial data, as well as some positive features of object-orientation where every physical or spatial object of the real world can be defined during software development. It is therefore imperative to investigate the practicality of a means to improve the representation of natural objects in 3D and to manage them in an object-oriented GIS.

Chapter 2 AN OVERVIEW OF 3D GIS DEVELOPMENT

The previous chapter has introduced the importance and some of the existing problems in 3D spatial data modelling and in developing an information system based on 3D spatial data. In this chapter, several types of two-dimensional (2D) GIS systems which are related to the development of 3D GIS will be further discussed. Some well established systems which are currently available in the market will be reviewed. Since data structures, data modelling and database management are important aspects of system development, all the discussions and system overview will focus on these aspects.

2.1 GIS Functions

Any GIS system should be able to provide information about geo spatial phenomena. Principally, the tasks or the functions of a GIS system are to: 1) capture, 2) structuring, 3) manipulation, 4) analysis, and 5) presentation (Raper and Maguire, 1992).

- *Capture.* Capturing is inputting spatial data to the system. Many different techniques and devices are available for both geometric and attribute data. The devices in frequent use for collecting spatial data can be classified as manual, semiautomatic or automatic, and the output either in vector or raster format. Detailed discussion on data capturing is not covered here.
- *Structure.* Structuring is a crucial stage in creating a spatial database using GIS. This is because it determines the range of functions which can be used for manipulation and analysis. Different system may have different structuring capabilities (simple or complex topology, relational or object-oriented).
- *Manipulate.* Among important manipulation operations are generalisation and transformation. Generalisation is applied for smoothing spatial data and it includes line smoothing, points filtering, etc. Transformation includes among others coordinate transformation to a specified map projection and scaling.
- *Analysis* is the core of a GIS system. It involves metric and topological operations on geometric and attribute data. Primarily, analysis in

GIS concerns operations on more than one set of data which generates new spatial information of the data. Terrain analysis (e.g. intervisibility), geometric computations (volume, area, etc), overlay, buffering, zoning are among typical analysis functions in GIS.

- *Presentation* is a final task in GIS. At this stage, all generated information or results will be presented in the form of maps, graphs, tables, reports, etc.

Ideally, a 3D GIS should have the same functions as a 2D GIS. However, such 3D systems are not available due to several impediments. The ensuing sections will discuss the challenges in 3D GIS development.

2.2 3D GIS

In this section, some problems and related issues in 3D GIS software development are reviewed and discussed. 3D GIS should be able to model, represent, manage, manipulate, analyse and support decisions based upon information associated with three-dimensional phenomena (Worboys, 1995). The definition of 3D GIS is very much the same as for 2D system. In GIS, 2D systems are common, widely used and able to handle most of the GIS tasks efficiently. The same kind of system, however, may not be able to handle 3D data if more advanced 3D applications are demanded (Raper and Kelk, 1991; Rongxing Li, 1994) such as representing the full length, width and nature of a borehole (some examples of 3D applications areas are listed in section 2.3). 3D GIS very much needs to generate information from such 3D data. Such a system is not just a simple extension by another dimension (i.e. the third dimension) on to 2D GIS. Adding this third dimension into existing 2D GIS needs a thorough investigation of many aspects of GIS including a different concept of modelling, representations and aspects of data structuring. Existing GIS packages are widely used and understood for handling, storing, manipulating and analysing 2D spatial data. Their capability and performance for 2D and for 2.5D data (that is also DTM) are generally accepted by the GIS community. A GIS package which can handle and manipulate 2D data and DTM cannot be considered as a 3D GIS system because DTM data is not real 3D spatial data. The third dimension of the DTM data only provides (often after interpolation) a surface attribute to features whose coordinates consist only of planimetric data or x, y coordinates. GIS software handling real 3D spatial data is rarely found. Although the problem has been addressed (as mentioned in chapter one) by several researchers such as Raper and Kelk

(1991), Cambray (1993), Rongxing Li (1994), Pilouk (1996), and Fritsch (1996), some further aspects particularly spatial data modelling using relational and OO techniques need to be investigated. This modelling issue will be addressed in later chapter.

2.3 Recent Progress Made on 3D GIS

Some recent research efforts by the GIS community has focussed on how to develop 3D systems; data structures and data models are major aspects of GIS system development. These efforts are summarised below.

Much previous work done on 3D data modelling concentrated on the use of voxel data structures (Jones, 1989). This particular approach does not address spatial modelling aspects (that is also topological aspect of the data); it is only useful for the reconstruction of 3D solid objects and for some basic geometric computations. Another problems with this data model is that it needs very large computer space and memory.

Carlson (1987) has proposed a model called the simplicial complex. He uses the term 0-simplex, 1-simplex, 2-simplex, and 3-simplex to denominate spatial objects of node, line, surface, and volume. His model can be extended to n-dimensions.

Cambray (1993) has proposed CAD models for 3D objects combined with DTM as a way to create 3D GIS, that is a combination of Constructive Solid Geometry (CSG) and Boundary representation (B-rep).

Other attempts to develop 3D GIS can be found in Kraus (1995), Fritsch and Schmidt (1995), and Pilouk (1996). These attempts were based on the TIN data structure to represent 3D terrain objects but no report exists on any related aspects of using OO techniques for modelling and data structure.

Data modelling and structuring of 3D spatial objects in GIS has not been as successfully achieved as in CAD (Li, 1994). Data modelling in GIS is not only concerned with the geometric and attribute aspects of the data, but also the topological relationships of the data. The topology of spatial data must be available so that the neighbouring and connectedness between objects can be determined. There are a number of mathematical possibilities for the determination of the topological description of objects. The information gained from the generated TIN neighbours is useful for further spatial analysis and applications. Topological relationships for linear objects as represented by TIN edges can be established. One edge is represented by a start node and an end node. From this edge topology, a chain of edges or arcs could be easily established. For TIN data, another approach is the simplicial complex developed by Carlson. A TIN's node is equivalent to 0-simplex, TIN's edge is equivalent to 1-simplex, a TIN surface (area) is equal to 2-simplex, and 3-simplex is equivalent to a 3D TIN (tetrahedron). The simplicial complex technique checks the consistency of generated TIN structures by Euler's equality formulae (see Carlson (1987) for a detailed discussion). An OO TIN approach is described in later chapter.

2.4 Commercially Available Systems and 3D GIS

There are few systems available in the market which can be categorised as a system which attempts to provide a solution for 3D representation and analysis. Four systems are chosen for detailed consideration. They were chosen because they constitute a large share of the GIS market and provide some 3D data processing functions. The systems are the 3D Analyst of ArcView (from Environmental System Research Institute or ESRI Inc.), Imagine VirtualGIS (from ERDAS Inc.), GeoMedia Terrain from Intergraph Inc. and PAMAP GIS Topographer. The following review is based on the available literature and Web-based product reviews.

2.4.1 ArcView 3D Analyst

The 3D Analyst (3DA) is one of the modules available in ArcView GIS. In ArcView these modules are known as extensions. The system's extensions and the main GIS module, that is the ArcView itself, is shown in Figure 2.1. ArcView is designed to provide stand alone and corporate wide (using client-server network connectivity) integration of spatial data (Maguire, 1999). The 3DA can be used to manipulate 3D data such as 3D surface generation, volume computation, draping for other raster images (such Landsat TM, SPOT, GeoSPOTV images, aerial photos or scanned maps), and other 3D surface analysis functions such as terrain intervisibility from one point to another (ESRI, 1997).

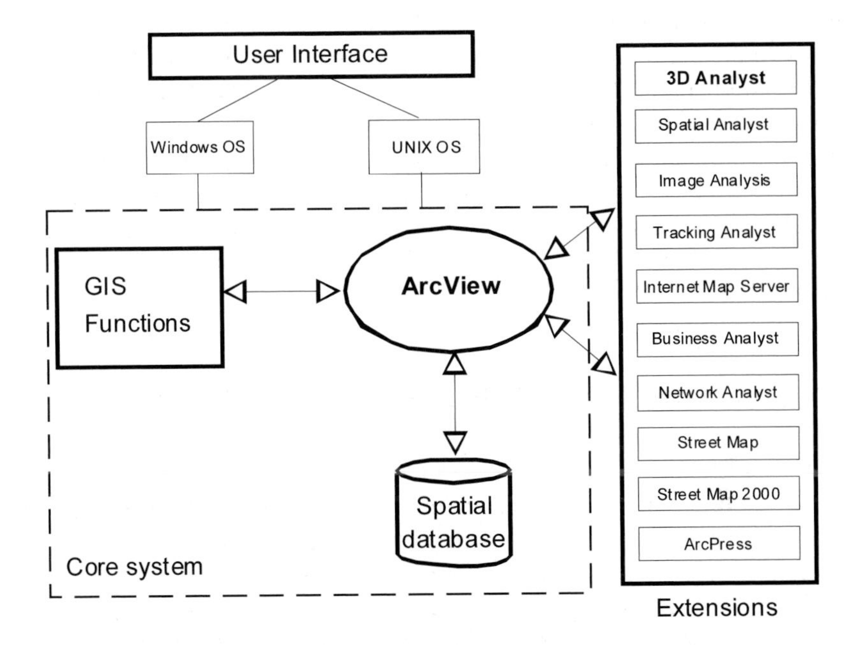

Fig. 2.1 The 3D Analyst (shown on top of the extension's box) within ArcView system

The system runs mainly on personal computers and accepts several operating system such Windows 95/98/2000 and Windows NT 4.0 as well as wide range of UNIX platforms (ESRI, 2000). The system works mainly with vector data. Even though raster files can be incorporated into 3DA, it is only for improving the display of vector data (e.g. by draping vector data with aerial photo images). (Raster files are and considerably for aspect of 2-D spatial data analysis.)

In summary, 3DA can be used to manipulate 3D data especially for visualization purposes. Thus, ArcView is very much a 2D GIS system, but 3DA supplies 3D visualization and display (e.g. of data with x, y, z coordinates). 3D GIS analysis is not achieved. It is worth noting, however, that 3DA supports triangular irregular network (TIN) data structure.

2.4.2 Imagine VirtualGIS

The Imagine system was originally developed for remote sensing and image processing tasks. Recently, the system has provided a module for GIS. The Imagine system is one of the GIS solutions developed by ERDAS Inc

(ERDAS, 2000). The GIS module is called VirtualGIS. It is a module that provides a three-dimensional visual analysis tools. The system has run under various computer systems ranging from personal computers to workstations such as DEC computers, IBM personal computers, Hewlett Packard, Sun Sparc and IBM RISC machines. Currently, the system works with operating systems such as Windows98/2000, Windows NT and various UNIX systems. It is a system which has an emphasis on dynamic visualisation and real-time display in the 3D display environment. Besides various and extensive 3-D visualizations, the system also provides fly-through capabilities (Limp, 1999). Figure 2.2 shows the system overview of the VirtualGIS with its core Imagine system.

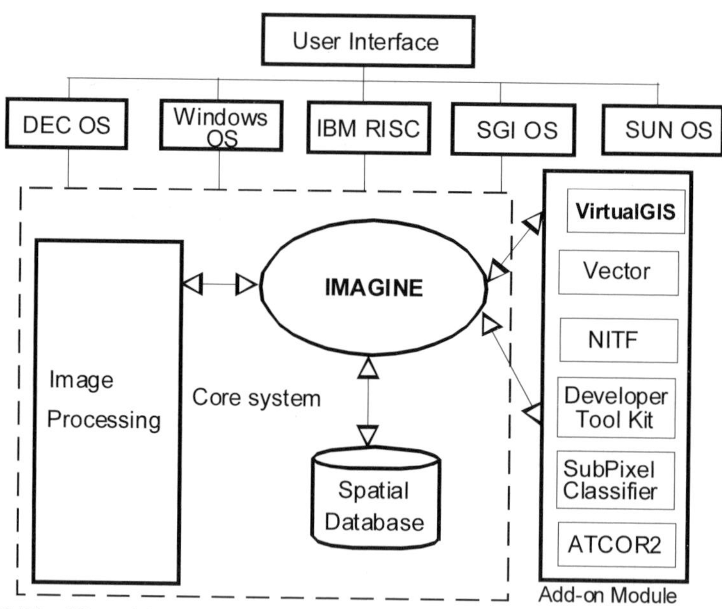

Fig. 2.2 The VirtualGIS component (shown on top of the Add-on module's box) in the Imagine system architecture.

As with 3DA, this system also centres around 3D visualization with true 3D GIS functions hardly available.

2.4.3 GeoMedia Terrain

GeoMedia Terrain is one of the subsystems that work under the GeoMedia GIS system developed by Integraph Inc. The system runs under the Windows operating systems (including NT 4.0 system). The Terrain system performs three major terrain tasks, namely, terrain analysis, terrain model generations, and fly-through (Integraph, 2000). In general, the Terrain serves as DTM module for the GeoMedia GIS as with other systems mentioned in the previous sections where true 3D GIS capabilities are hardly offered by software vendors. Figure 2.3 shows the Terrain subsystem within the GeoMedia core system.

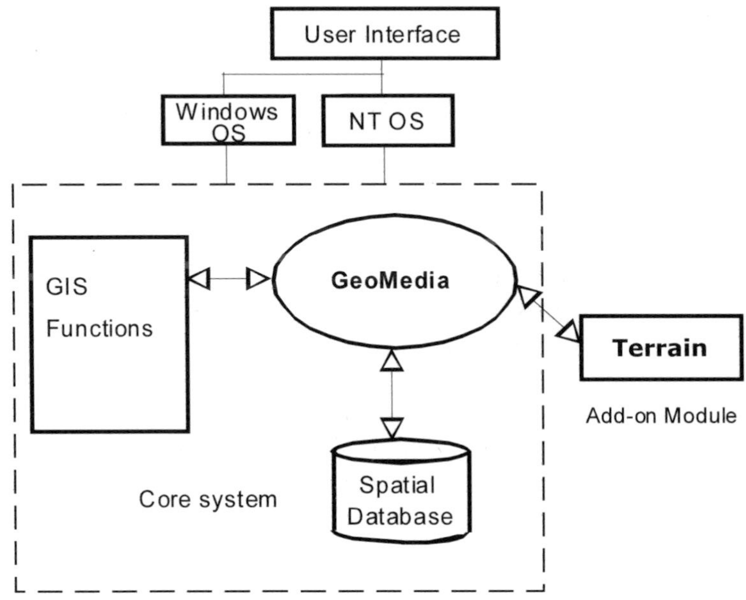

Fig. 2.3 The Terrain component within the GeoMedia system

2.4.4 PAMAP GIS Topographer

This GIS system is one of PCI Geomatics Inc.'s products. It runs under Windows95/98 and NT operating systems. PAMAP GIS is a raster and vector system (Geomatics, 2000). Besides its 2D GIS functions, the system has a module for handling 3D data, called Topographer as depicted in Figure 2.4. Four main GIS modules are offered, they are Mapper, Modeller, Networker and Analyser which form the core system. For 2D data handling, the system performs GIS tasks as in other systems mentioned earlier. For 3D data, most of the 3D functions in the Topographer work as by any DTM packages, for example terrain surface generation, terrain surfaces analysis (e.g. calculation of area, volume) and 3D visualisation (such as perspective viewing). This system also focuses on 3D display of terrain data.

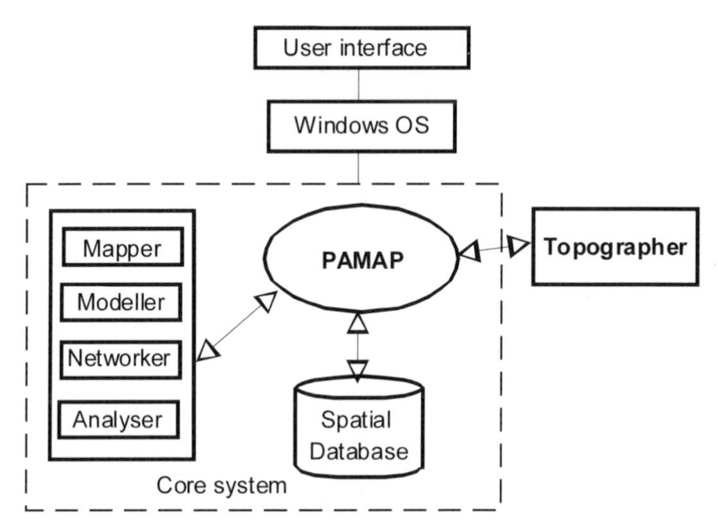

Fig. 2.4 The Topographer within the PAMAP GIS system

In summary, all the systems discussed here show little provision of 3D GIS functionality even though most of them can handle 3D data efficiently in the aspect of 3D visualization. A fully integrated 3D GIS solution has yet to be offered by any general purpose GIS vendor.

There are, however, few prototype 3D GIS systems and one of them is developed by Fraunhofer Institute, Germany. This system utilises a CAD modeller which can generate 3D objects (such as buildings) on top of the terrain (Rimscha, 1997). Another prototype system which was developed by an Austrian company Grintec has tested the system within urban objects. The system, called GO-3DM also uses CAD and DTM for the management of the city of Graz's 3D objects (mainly buildings) as reported by Rimscha. Despite some exciting developments in 3D visualization and the possibility of incorporating them within GIS, true 3D GIS solutions remain to be realised. This indicates that 3D GIS has far from arrived and needs further investigations.

2.5 Why is 3D GIS Difficult to Realise?

The difficulties in realising 3D GIS or 3D geo-spatial systems stem from:

- Data structures: although there are several data structures available for the 2.5 D and 3D data, each of them has its own strong and weak points in representing spatial objects; and
- Data models: spatial data can be modelled in different ways. Any model should be able to describe relationships between data in such a way that information can be generated from them.

This monograph attempts to address these two major issues by investigating the possible uses of several data structures (including some 2D structures), the construction of these data structures, the utilisation of these structures in spatial modelling, the development of a database from the spatial data, and the implementation of them in the form of a software package which can be seen as a component of GIS.

2.6 Discussion

From the foregoing discussions the problem of data structuring and data modelling for 3D data in analytical GIS environment remains unsolved. The only near solutions offered concentrate on the visualisation aspect as indicated in section 2.4. This gap of GIS functionality needs to be investigated. The effort carried out in this research work focuses on the spatial data structuring and data modelling with emphasis on developing a software which will contribute towards 3D GIS. To do this, several existing pertinent data structures are investigated which can handle 2D as well as 3D data. This effort is realised in the form of software development which covers aspects of data structuring, relevant algorithms development, data modelling using object-oriented technique and a simple front-end OO interface.

Chapter 3 2D AND 3D SPATIAL DATA REPRESENTATIONS

In the geoinformation domain, two-dimensional (2D) and three-dimensional (3D) spatial data are commonly available. There is no doubt that 2D data are utilised much more than 3D. This situation is attributable to several factors including difficulties in 3D data structuring, particularly topological data structuring (Raper, 1992; Li, 1994). These problems need to be investigated so that the feasibility of having a system capable of handling both 2D and 3D data types can be assessed. This chapter focusses on the subject of spatial data representation in an attempt to contribute to an understanding of how spatial data could be utilised for a geoinformation system. The chapter aims to review some of the pertinent spatial data representations and adopt suitable structures for a geoinformation system capable of handling 2D and 3D spatial data.

3.1 Introduction

Geospatial data can be represented in three clearly distinct Euclidean dimensional contexts: 2D defines location by measurements on the XY axes; 2.5D defines location in 2D space with a dimensional attribute value attached to the XY location, for instance elevation above datum (Z coordinate) may act as the attribute value; 3D defines location extending through 3D space defined by X, Y, and Z axes (Raper, 1992). These locations position real-world spatial objects which could be regular or irregular in shape. Man-made objects, e.g. buildings are examples of regular objects while terrain surfaces, forests, sea floors, trees and algal blooms are examples of irregular objects. All real world objects are three-dimensional (3D). How can objects be represented in a system where information regarding the state, behaviour, and the topological relationships of the objects with their neighbours can be elegantly retrieved? There exists no straightforward answer to this question. In GIS, spatial objects are represented in the form of points, lines, and surfaces. These primitives work well for two-dimensional (2D) objects as described by Peucker and Chrisman (1975), but these authors did not consider 3D objects at all. As the demand from GIS applications in the 3D environment increases, the basic forms (e.g. single z-value for an xy location) of data representation are no longer adequate (Raper and Kelk, 1991). As a result, work has emerged attempting to solve the problem, but much has focussed on regular objects (Cambray, 1993; Bric, 1994) such as buildings, houses, etc.

Representing non-regular objects needs different data representations so that the general shape of objects can be represented. The following sections look into several existing types of representation that can be used for 2D and 3D data.

3.2 Classes of Object Representations

As an initial classification, object representations may be described as surface-based and volume-based (Li, 1994). Li called an object a surface-based representation if the object was represented by surface primitives. It is volume-based if an object's interior is described by solid information. Figure 3.1 shows the two categories of spatial object representations.

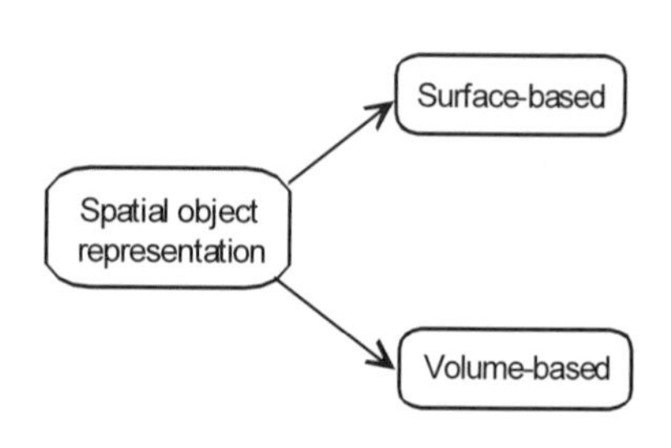

Fig. 3.1 The two categories of spatial object representations

The surface-based representations are: grid, shape model, facet model, and boundary representation (b-rep). The volume-based representations are: 3D array, octree, constructive solid geometry (CSG) and 3D TIN (or TEN). Some of these representations are common in computer-aided design (CAD) systems but not in GIS. Figure 3.8 illustrates the list of surface-based representations with their basic elements. Figure 3.15 illustrates the list of volume-based representation with their basic elements.

The following sections describe the surface-based representations.

3.2.1 Grid

A grid is a widely used method for surface representation in GIS, digital mapping and digital terrain modelling (DTM). It is a structure that specifies height values at regular locations (see Figure 3.2). Many DTMs and terrain surface packages are based on this representation for generating surfaces as reported in Petrie and Kennie (1990). This structure has several advantages; it is simple to generate, and topology information (in terms of positions) is implicitly defined (Peucker, 1978). (In this structure, the topology of grid points can be easily determined since each grid point is relative to other points). The structure may be considered as an array structure in computer programming. Each array element represents the XY locations of the grid.

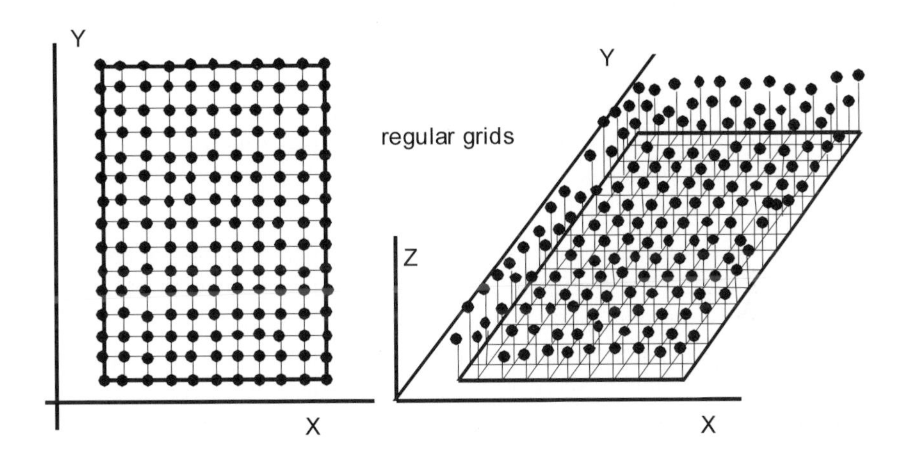

Fig. 3.2 Grid representation of surfaces (orthogonal and perspective views)

The relative positions (i.e. the topology or the neighbouring points) of the grid points are easily defined, and they could be regular or irregular. Although excellent terrain surfaces can be derived with this structure, it is not helpful for surfaces of multiple heights, e.g. vertical walls or overhangs (Heitzinger and Pfeifer, 1996). In fact, this is one of the major drawbacks of the structure. Although it can represent surface points well, incorporating other terrain objects or terrain breaklines such as linear, polygonal, and even more complex features needs extra geometric computations and interpolations with the grid points. A better model than a grid is thus desirable.

3.2.2 Shape Model

A shape model describes an object surface by using surface derivatives (e.g. slopes) of surface points (Rongxing Li, 1994) as shown in Figure 3.3. In this model, each grid point has slope value instead of Z value. With known slopes, a normal vector of a grid point can be defined and used to determine the shape of the surface. An experiment reported by Rongxing

Li (1994) showed that the structure has an application in surface model reconstruction especially for sea bed surface mapping.

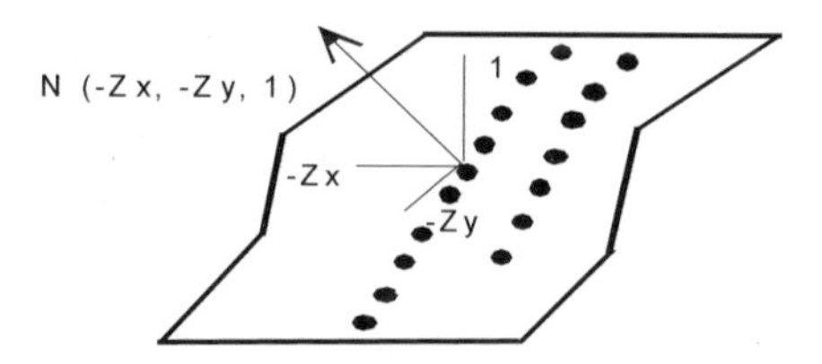

Fig. 3.3 An example surface determination using shape model (after Rongxing Li, 1994)

Although the technique can be used for sea bed surface mapping, the usage of such technique for on-surface terrain mapping may need to be investigated especially on the aspect of data acquisition. In this technique, slopes of grid points are determined by usimg image processing technique (detailed computation technique can be found in Rongxing Li (1994)). This model works with regular or irregular XY locations as with the grid approach, and thus it has the same surface mapping capability as the grid (discussed in section 3.2.1).

3.2.3 Facet Model

A facet model describes an object's surface by planar surface cells which can be of different shapes and sizes. One of the most popular facet models uses triangle facets, sometimes known as a triangular irregular network (TIN). A surface can be described by a network of triangle facets. Each facet consists of three triangle nodes which have a set of x, y, z coordinates for each node (see Figure 3.4).

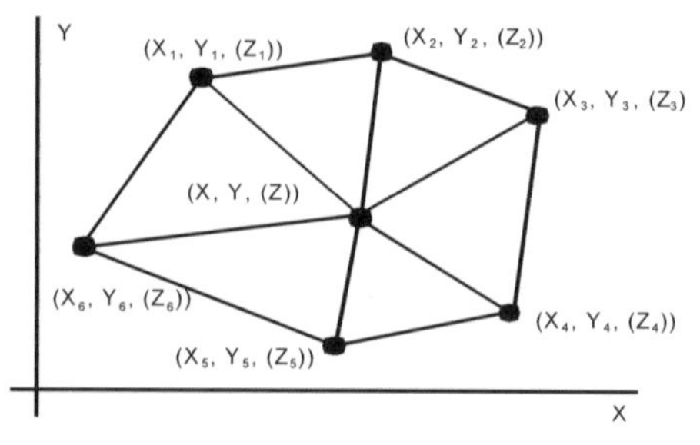

Fig. 3.4 2D TIN model

Figure 3.5 shows a distribution of points on the real world. The triangle structure is widely used in DTM and other terrain surface software mainly because of its structural stability and terrain feature adaptability (Midtbø, 1996), data interpolation simplicity (Abdul-Rahman, 1992) and also for object visualization (Kraak, 1992). Triangles or TINs as illustrated in Figure 3.6 can be constructed in the raster or the vector domain, where most of the triangulations techniques are based on the Delaunay triangulations.

Briefly, one way to generate triangles in the raster domain is first by rasterising all surface points. (These rasterised points are sometimes known as kernel points in raster data processing.) That is by using a distance transformation (DT) technique to each kernel point. DT calculates the distances of each point to the neighbouring points. Each kernel point has its dual image that is a Voronoi polygon of surface points. Then, from three neighbouring Voronoi polygons, a Delaunay triangle can be established (i.e. three points represent one triangle). Thus, a set of triangles can be established from a set of Voronoi polygons.

The shapes and sizes of the triangles vary, depending on the original distribution of the data sets. One of the advantages of this representation is that the original observation data are preserved, that is, all surface points are used for surface representation. Figure 3.6 shows an example of TINs generated from random distributed points. The points were acquired using ground land survey technique. Figure 3.6 illustrates that terrain surfaces in the form of random distributed points are well represented by this planar facet representation.

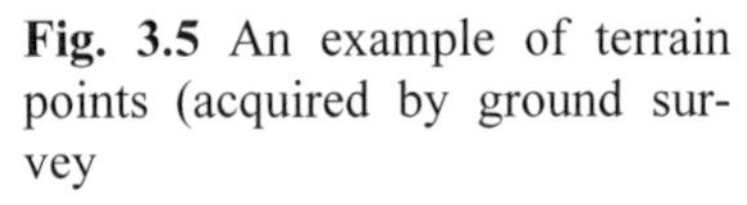
Fig. 3.5 An example of terrain points (acquired by ground survey

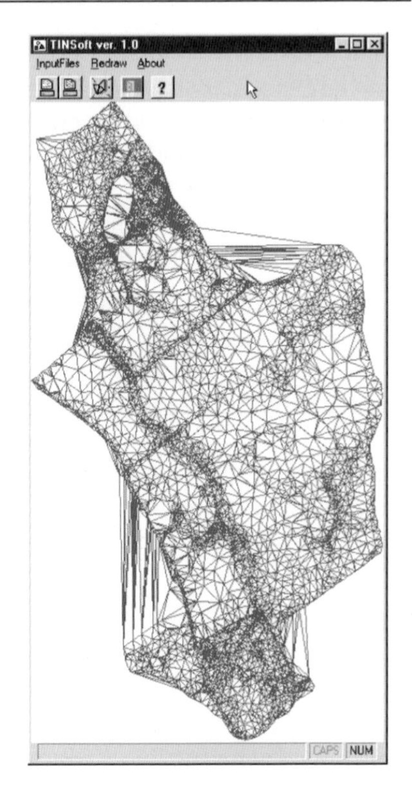

Fig. 3.6 An example of TINs facet representation of terrain surfaces for points as depicted in Fig. 3.5

TIN facets using digitized contours and photogrammetrically acquired data sets were also generated and are presented towards the end of this book.

3.2.4 Boundary Representation (B-rep)

Boundary representation (B-rep) represents an object by a combination of predefined primitives of point, edge, face, and volume. Examples of point elements are individual points, contour points, and other auxiliary points which approximate a curve or a face. Examples of edges are straight lines, arcs, and also circles. Examples of faces are polygon planes and other spatial object faces such as arced faces, cone and cylinder faces. Volumes are an extension of surface elements for representing volume characteristics in B-rep. They may consist of boxes, cylinders, cones, and other combinations. To represent an object by this model, an element of B-rep needs to have a geometric data element, an identification code of element and its relationship to other elements (Rongxing Li, 1994). Figure 3.7 shows a simple B-rep representation of a polygon object. Here, the key element of constructing an object is primitive combinations, i.e. a combination of points to form an edge, combination of edges to form a planar surface.

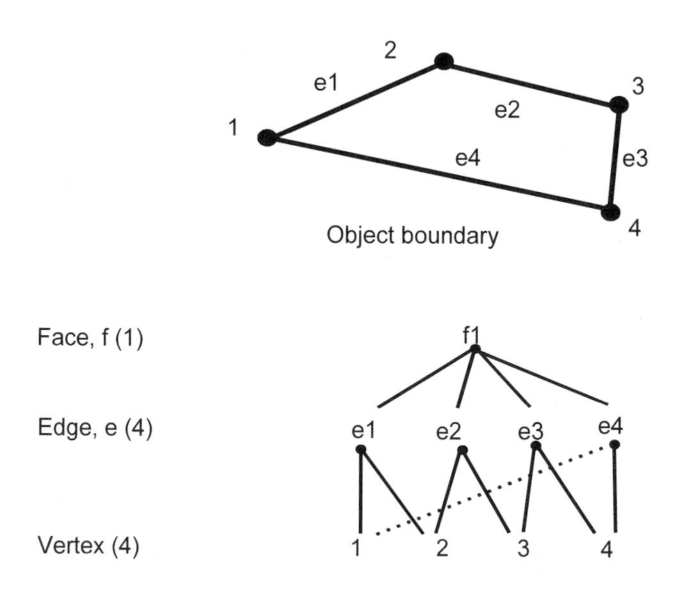

Fig. 3.7 Planar polygon representation of B-rep

For non-planar surface, smooth surface functions such as a Bezier surface or B-spline functions could be incorporated in the surface generation, and this normally involves a considerable amount of geometric and complex computations. Although B-rep is popular in a computer-aided design/computer-aided manufacturing (CAD/CAM), due to computational complexity and inefficient Boolean operations, it has been suggested that B-rep is only suitable for regular and planar objects (Mäntylä, 1988; Rongxing Li, 1994). In GIS, the use of B-rep for representing spatial objects is very limited because the model needs to be modified in such a way that the three fundamental spatial data elements, i.e. geometric, attribute, and object identification data can be stored together with the related topological data. Figure 3.8 illustrates a summary of the surface-based representation of 2D objects.

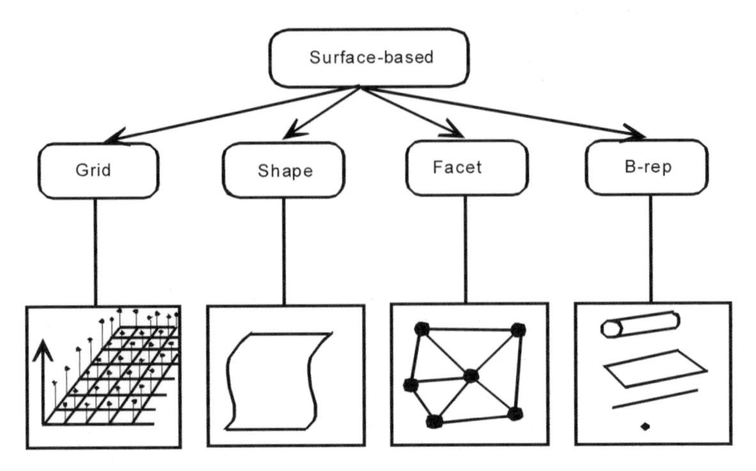

Fig. 3.8 Examples of surface-based representations

The following sections describe the volume representations of 3D objects.

3.2.5 3D Array

3D array is perhaps the most simple data structure in the 3D domain. The structure is easy to understand and to implement, but it may not be efficient for some tasks. For example, if many array elements are occupied with the same values, it creates a huge but unnecessary demand for computer storage space and memory. Thus, it is less suitable for representing objects at higher resolution since storage and memory increase with higher resolution.

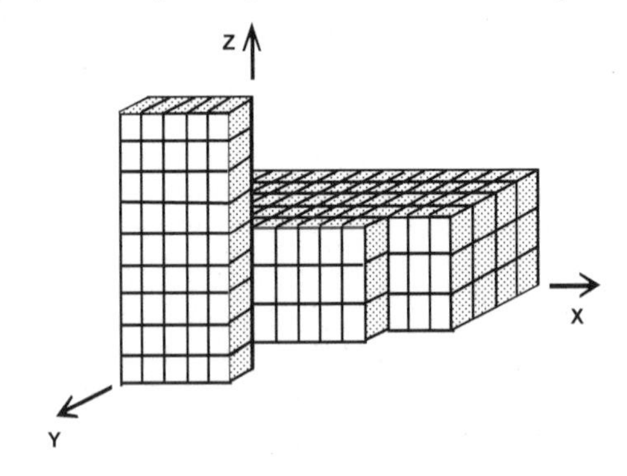

Fig. 3.9 An example of 3D array representation for solid object

In the 3D array shown in Figure 3.9, the size of the array elements is equal and each occupies the same amount of computer space although the voxel size can be specified and controlled by a program. 3D array needs a huge computing power and this is one of the reasons why this kind of representation is seldom used in practice (Feng Dong, 1996).

A much better way of representing 3D objects is by varying the size of the voxel that is the octree technique.

3.2.6 Octree

The term octree refers to a hierarchical data structure that specifies the occupancy of cubic regions of the object space. These cubic regions are often called voxels. This representation has been used extensively in image processing and computer graphics (Samet, 1984). It is a data structure that describes how the objects in a scene are distributed throughout the three-dimensional space occupied by the scene. It is simply a three-dimensional generation of a quadtree. Conceptually, the area of interest is enclosed by a cube represented by voxels (Mark and Cebrian, 1986). As in the quadtree structure, the octree is based on recursive decomposition, and can be used to encode 3D objects (Meagher, 1982; Jones, 1989; Chen, 1991; Brunet, 1992; Rongxing Li, 1994; Feng Dong, 1996). In the octree approach, each node is terminal or has eight descendants. The tree divides the space of the universe into cubes which are inside or outside the object. The root of the tree represents the universe, a cube with an edge of length 2^n. This cube is divided into eight identical cubes, called octants with an edge length of 2n-1. Each octant is represented by one of the eight descendants of the root. If an octant is partially full of solid, it is termed a "grey node", and it is divided into another eight identical cubes which are represented as descendants of the octants in question. This process is repeated recursively until octants are obtained which are either totally inside the solid ("black nodes") or totally outside it ("white nodes") (see Figure 3.10). A minimum octant size (i.e. a threshold) which determines the number of subdivisions of if the octants is one of the important factors in octree processing. Meagher (1982) also reported that one of the advantages of the octree approach is its simplicity for Boolean operation and visualization rendering algorithms, but it has a drawback in terms of storage space.

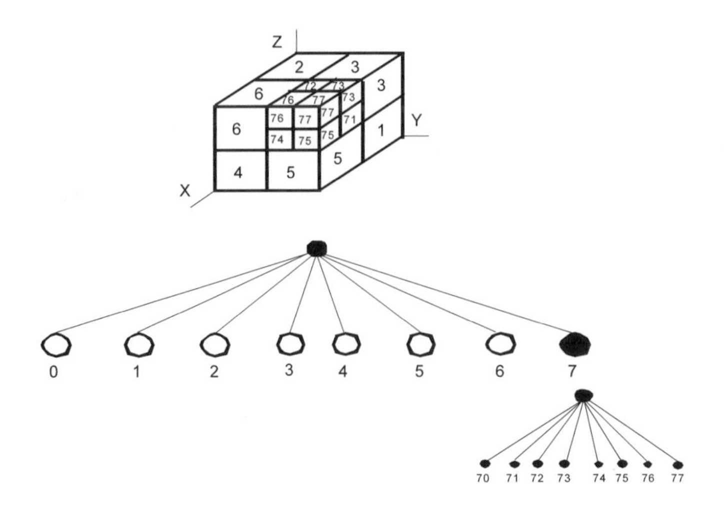

Fig. 3.10 An example of Octree representation of object

To represent detailed objects, a large amount of storage space and more processing power are needed. One way to overcome this problem is by using an octree model called the 'vector octree' as proposed by Samet (1984) and also reported in Jones (1989). In the vector octree, three types of octree nodes are introduced, namely, face node, edge node and vertex node. These extra nodes are used to represent object surfaces, and reduce the degree of subdivision. They, thus require less storage. Rongxing Li (1994) also reported that the octree approach is very efficient in spatial analysis, Boolean operations, and database management because of their hierarchical data structure.

3.2.7 Constructive Solid Geometry (CSG)

Constructive solid geometry (CSG) represents an object by a combination of predefined simple primitives called geometric primitives (see Figure 3.11). The examples of primitives are spheres, cubes, cylinders, cones, or rectangular solid, and they are combined using Boolean set operators and linear transformations as discussed in Mäntylä (1988). CSG is commonly used in solid modelling such as CAD/CAM because object creation can be completed interactively with a simple modelling language (Raper, 1989). This representation is also widely used in engineering and architectural visualization because constructing primitives or solid geometries is usually straightforward (Feng Dong, 1996).

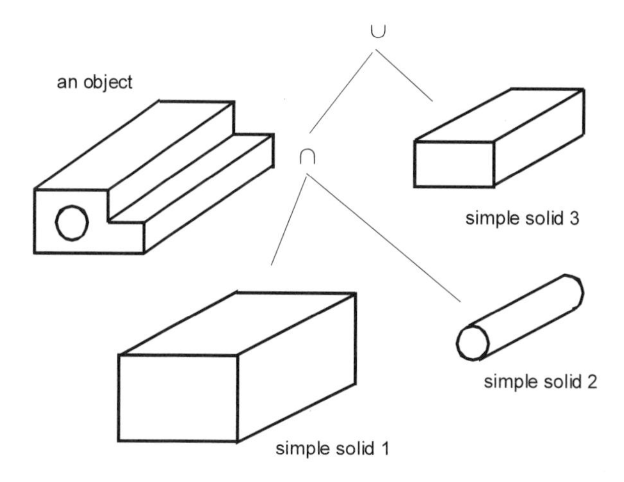

Fig. 3.11 Simple objects from CSG simple primitive solids

The primitives of CSG are regularly shaped volumetric instances and can be combined by using geometric transformation and Boolean operations. The geometric transformations normally involve translation, rotation and scaling, and Boolean operations normally involve union, intersection and subtraction (or differencing). The storage space of CSG increases as the number of primitives increases (Samet, 1990). Previous research suggested that CSG is only suitable for describing regular shaped objects (Cambray, 1993; Rongxing Li, 1994) because the primitive combinations of regular objects to form irregularly shaped volumetric instances needs considerable computing effort. Thus, CSG is currently considered not well suited for irregular objects.

3.2.8 3D TIN (Tetrahedral network, TEN)

Basically 3D TIN is an extension of 2D TIN, sometimes called TEN (short for a Tetrahedral Network). An object is described by connected but not overlapping tetrahedra. Similar to 2D TIN, TEN has many advantages in manipulation, display and analysis. A TEN is made of tetrahedra of four vertices, six edges, and four faces. This representation has been considered a useful data structure in earth sciences by researchers for some time (Raper and Kelk, 1991). It can be generated using the same techniques as for 2D TIN. If we build a 2D TIN from 2D Voronoi processing, then 2D Voronoi processing can be extended to 3D. 3D TIN can be derived from 3D Voronoi polyhedrons (Qingquan and Deren, 1996). Other techniques of generating TEN can be found in Midtbø (1996).

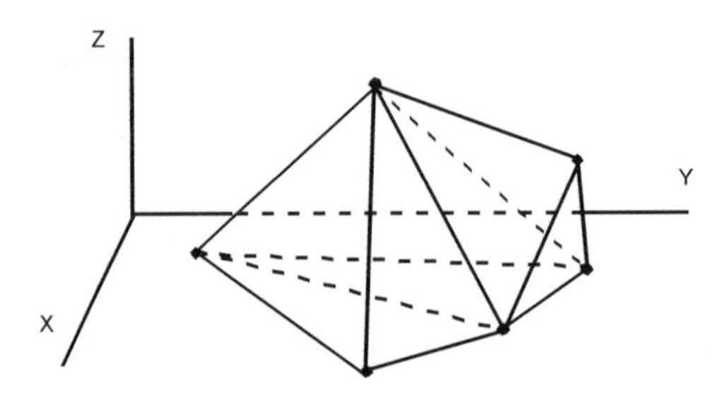

Fig. 3.12 An example of 3D TIN (TEN) model

Previous work has pointed out that TENs have several advantages over other solid structures. The advantages are that it is the simplest data structure that can be reduced to point, line, area and volume (solid) representations; it supports fast topological processing, and it is also convenient for rapid visualization. However, work on tetrahedra for GIS is very limited. A screen shot illustration in Figure 3.14 shows an example of 3D TINs generated from simulated boreholes datasets of Figure 3.13. Each borehole has several height locations with the same XY coordinates.

This particular example indicates that TEN can be used to manipulate underground 3D objects such as boreholes. Volume computation of lithologies between boreholes is one of the possible 3D modelling tasks. Other applications such as iso-surface generation are also possible as often demanded in Earth Science applications.

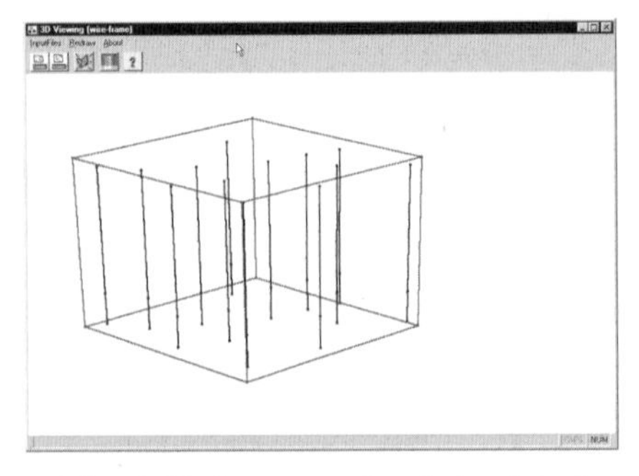

Fig. 3.13 An example of simulated boreholes

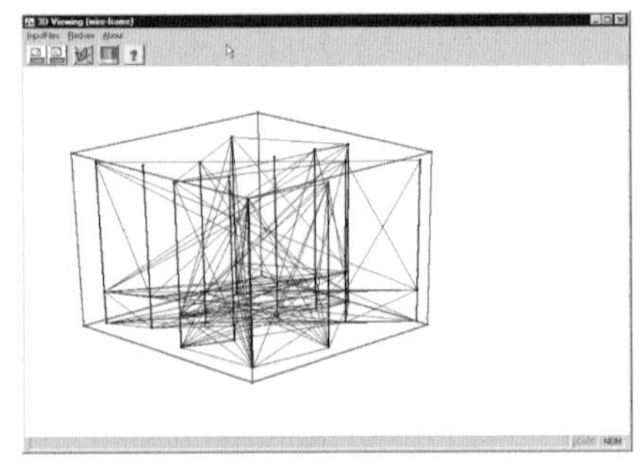

Fig. 3.14 An example of 3D TIN representation for the boreholes

Figure 3.15 illustrates the summary of the volume-based representations that can be used for 3D objects.

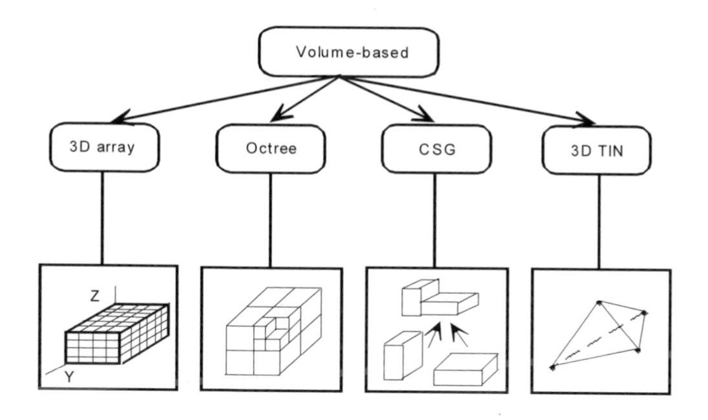

Fig. 3.15 Examples of the volume-based representations

3.3 GIS Applicability of the Representations

Based on the previous section, it can be seen that surface-based representations describe the geometric characteristics of objects by surface entities. Grids, shape models, and facet models are suitable for describing irregular object surfaces, while the B-rep model is more for exact surface geometry of regular shapes. For volume-based representation, 3D array, octree, and 3D TIN (or TEN) can be used for irregular objects. The 3D TIN and octree models can be used for volume objects. Octrees are an approximate representation, therefore a very detailed representation of objects is hard to achieve; with octrees, storage space increases rapidly when the resolution increases. Although computer storage is less of an issue these days, there is little evidence of success in using the model for spatial data representation despite the convenience in volume computation and visualization as reported by Turner (1992a).

As for TEN (3D TIN), it is suggested that the model is able to represent objects accurately, describe complicated spatial topological relations and maintain the original observations (Qingquan Li and Deren Li, 1996). Thus, we can initially assume that irregular objects can best be represented by 3D TIN and octrees. To make a choice between these two representations for irregular objects, however, is a difficult task. The next section attempts to define some means for selecting the most appropriate representation.

3.4 The Selection Criteria

The discussion and summaries of the previous sections have shown that two representations stood out as the most suitable for irregular objects: TEN (or 3D TIN) and octree. Between these however, which is the most appropriate one? Two major items should be considered when selecting the best representation:

- The ability to represent (or to be converted to) object primitives, e.g. points, lines, surfaces, and areas.
- The ability to integrate topological and attribute so that geospatial database queries and data retrieval can be performed.

The following section describes the association of these two properties within the tetrahedral (3D TIN) and octree approaches.

3.4.1 Representation of Object Primitives

In the real world, the points, lines and areal features with which we have traditionally populated our cartographic databases do not exist. Furthermore, surfaces as they are represented in spatial databases are a reduced description of real world objects - being a representation of a part of the object described locationally with respect to a surface such as the mean sea level or a spheroid. In reality, all objects, as we perceive them and should use them at the level of detail supported by a 'typical' GIS, are irregular and three-dimensional, having more or less well defined bounding surfaces separating them from other such irregular three-dimensional objects; they are not points, lines, areas and surfaces.

Given that real world objects are all irregular and three dimensional and can all be adequately represented using either the TEN or octree approaches, for reasons of efficiency or convenience, the chosen data may be processed in a more primitive form (i.e. as points, lines, areas or surfaces). A GIS processing example is route selection. Thus, a consideration is needed whether either or both TEN and octree representations can be reduced to the object primitives and which representation can more easily be reduced to the required object primitives.

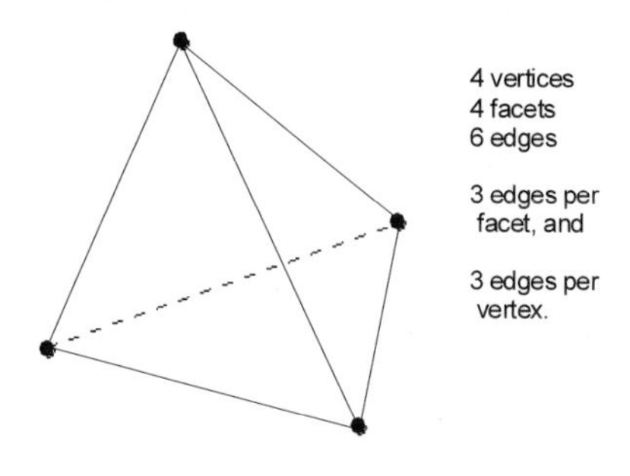

Fig. 3.16 The tetrahedron (3D TIN) primitives

Figure 3.16 shows a tetrahedron, the fundamental building unit of the TEN approach. For purposes of illustration, let us consider a city, data relating to which is to be processed as if the city were a point entity. Within a city are buildings, streets and other utilities, trees, street furniture, waterways, etc. Each of these real world objects can be represented using the TEN approach; each tetrahedron's description includes vertices and attributes. All objects belonging to the city will be appropriately attributed and retrievable. The mean of the x, y coordinates of all the vertices of these retrieved objects can provide x, y coordinate pairs to allow representation of the city as a two-dimensional point feature.

Considering a particular street represented by tetrahedra, for each tetrahedron at least one facet will represent the street surface, and the vertices of this facet will be points along the street. A centre-lining procedure can generate a line from the set of vertices from all the surface facets of the street's tetrahedra.

Considering an object as a piece of undeveloped land within a city represented using the TEN approach, some tetrahedron edges will be the edges between two 'undeveloped land' tetrahedra, and some will be the edges between undeveloped and land of another category. Those edges representing change in categories are the edges of the undeveloped land, and the x, y coordinates of their vertices represent the bounding polygon (or 'area') of the undeveloped land.

Finally, considering the surface of the city itself, if this has been described by a series of tetrahedra, as with the street, some facets will be surface facets and their vertices describe an irregular DTM. It is possible that the coordinate system used to describe vertices' locations may not be appropriate for the DTM (e.g. with respect to an inappropriate datum). An appropriate coordinate transformation will need to be introduced.

Octree works with 3D raster data sets. It is therefore the case that all object entities have to be converted into 3D raster for further processing. These objects then need to be decomposed into point, line, surface, and solid primitives if they are to be used in a GIS, for example. A number of authors have reported on the use of octree for GIS, but most of them have focussed on visualization and volume computation tasks (Chen, 1991; Mark and Cebrian, 1986; Meagher, 1982). Work on octrees with the related aspect of spatial data modelling is less reported. Most research work on octrees was for solid modelling and visualization purposes as reported in Turner (1992b).

The discussions thus far have shown that TEN representation is a more promising model for 3D spatial objects than octree.

3.4.2 Topology of Spatial Objects: Simplexes and Complexes

In GIS, besides geometric and attribute data, topology has a vital role in spatial information system. Topology is used to determine the connection relationships of objects in space. For example, in the case of a point object, one may need to know its relationship with neighbouring objects (where it could be with points, lines, areas, or solid objects). The same holds for lines, areas, and solid objects. A number of researchers have looked into this topological problem, including Frank and Kuhn (1986) and Worboys (1995). They all use the term complex and simplex for describing the topological relationships of planar objects. In the 2D case, triangular irregular network structures can be regarded as simplicial complexes in a Euclidean plane. Here, a 0-simplex is the set of a single point in the Euclidean plane. A 1-simplex is a straight line segment. A 2-simplex is a set of all the points on the boundary and in the interior of a triangle whose vertices are not collinear. These simplices are well represented in the facet model of representation (see section 3.2.3) where a TIN node is topologically equivalent to 0-simplex, the edge of a TIN is topologically equivalent to 1-simplex, and a TIN area (surface) is topologically equivalent to 2-simplex. Since this simplicial complex theory is extendable to n-dimension, then we could also represent TEN primitives using the same principle. That is a 3-simplex is a volume which is a tetrahedron (see Figure 3.17).

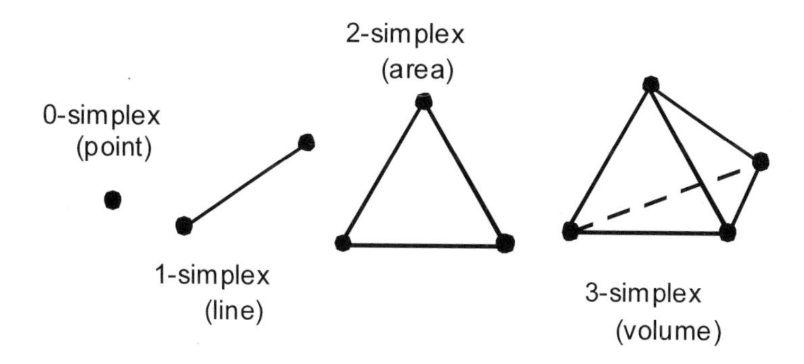

Fig. 3.17 Example of simplices (0, 1, 2, and 3)

Simplices are the building blocks of a larger structures, the simplicial complexes. Complexes are built from simplices. If we recall the TIN representation (see Figure 3.4), a simplicial complex can be formed (i.e. two-dimensional complexes). This concept of simplicial complex provides a sound framework for analysis of the topology of a mixture of points and edges in a plane and is workable for the TIN representation of spatial objects (both 2D and 3D) as cited in Worboys (1995).

3.5 Vector and Raster Representation

Geoinformation data may come in vector form, raster form, or in both forms. Spatial objects are said to be in vector form if they are represented by one of the basic discrete entities such as points, lines, and areas (polygons) which are spatially referenced by a Cartesian coordinate system (Burrough and McDonnell, 1998). The same spatial object entities can be represented in raster form if they can be decomposed into pixels. Each pixel is referenced by row and column positions. Representing spatial objects as raster or vector has its advantages and disadvantages. Vector representation easily offers better accuracy than raster representation because entities are represented by exact coordinates in space and do not have their locations generalized to a pixel. Thus, raster gives more approximate locations for the represented entities. Further comparisons of these two representations such as based on handling topology is explicitly described in the vector form and therefore this is good for tasks such as network analysis. However, geometric data processing such as coordinate transformation is difficult (requiring resampling) in raster but easy to perform in vector form

(Burrough and McDonnell, 1998). A further debate on these two representations can also be found in Antenucci *et al* (1991) and Chou (1996).

The choice between the two representations depends on factors such as processing speed, level of difficulty, etc. In this research, we used the raster form as a means of data processing for 2D and 3D TIN model construction and also for related data structuring. That is due to the simplicity of raster data processing. The discussion in section 3.2.3 indicates that TINs could be constructed using rasterised datasets. The simplicity of raster data processing for the two object representations is also examined and described in chapter eight.

3.6 Summary

From the foregoing discussion of 2D object representations, 2D TIN has been shown to have several advantages over the other models of the same category (i.e. the grid, shape, and the B-rep.). The model's promise relies on the fact that it can be used to construct a generic data structure (including topological relationships). Other models such as grid, shape, and B-rep require further structure modifications before they can be used, and thus they lead to expensive modelling in the digital environment.

Since 2D TIN can be extended to 3D TIN and have similar geometric properties, 3D TIN can represent 3D spatial objects. An important property of the model (or the structure) is that simple object primitives are aggregatable into a larger object. The aggregation of features into more complex features is perhaps the most important feature in spatial data modelling. Models other than 3D TIN have some drawbacks in this task, in that they require a huge computing effort. For example, real world spatial objects are complex in nature and it is obvious that tremendous decomposition operations are involved if one deals with them as octrees. Although the octree approach is widely used in the solid geometry visualization community, difficulty in spatial data structuring and the related topology entails limited practicality in GIS.

The pertinent spatial object representations have been described and TINs (2D and 3D) have been identified as the most appropriate representations for the 2D and 3D spatial objects. Thus, these structures become the major focus for the development of a geo information system in this research.

The modelling and other relevant fundamental aspects of the geoinformation system will be discussed in the next chapter.

Chapter 4 THE FUNDAMENTALS OF GEO-SPATIAL MODELLING

In general, a GIS can be considered to have several components such as spatial, graphical, numerical, and textual components (Worboys, 1995). These system components have several important building blocks such as data modelling, data structures, and types of applications. However, Molenaar (1996a) argues that it is the process of spatial data modelling alone which leads to the development of a complete geoinformation system. This chapter introduces the fundamental concepts of spatial data modelling and GIS. The concept of modelling spatial data will be investigated, as well as the construction, manipulation and management of spatial data within the development of a GIS system. In particular, concepts such as spatial data, modelling of spatial data, construction, manipulation and management of spatial data in the domain of the triangular irregular networks (TINs) data structure are the foci of this chapter. The aim is to describe major processes and steps involved in the development of a system which is based on TIN spatial data. Although this system is far from complete (since it does not contain, for example a temporal aspect), most of the major components and the related building blocks for the system are considered. (Relevant temporal aspects of GIS are addressed in Langran (1992) and Wachowicz (1999)).

The layout of the proposed TIN-based system is presented at the end of this chapter, following the discussions on spatial data, spatial data modelling, data structuring, database models and the related database management systems (DBMS).

This chapter also reviews various concepts fundamental to spatial modelling and specifically related to geo-information. The aim is to outline the theoretical bases and fundamental concepts necessary for the design of a geo-spatial model. Since spatial theory is a relatively young discipline developed from a combination of many branches of mathematics and computer science, the terminology found in the literature can be confusing. In this chapter, the terms used in this monograph are clarified. Five important components of spatial model and phases of modelling are defined. The emphasis is placed on the conceptual and logical designs of spatial models. Mathematical concepts concerning space, objects, and their interrelationships are taken as the foundation of the conceptual design of a spatial model. The concept of a simplicial complex and the theory of graphs are chosen as methods of representing objects and their interrelationships in the model. Existing spatial models are taken as examples to show the lines of further development. Relational and object-oriented approaches are considered important for the logical design of a spatial model.

4.1 Spatial Data

Figure 4.1 shows the basic components of spatial data. Principally, there are three spatial data components that need to be stored for GIS data: geometric data, thematic data, and a link identification (ID) for the geometric and the thematic component. The illustration in Figure 4.1 shows the link between the geometric component (which deals with the location of the data by means, for example, of a reference coordinate system) and the thematic component (it provides the attribute values of the data, e.g. names, and other identifiers (IDs) of the data). Object or feature needs to be geometrically and thematically described (Longley *et al.*, 1999; Laurini and Thompson, 1991). The basic components of spatial data (TINs) can be used to describe real world terrain objects, whether natural or man-made; thus we have TIN-based spatial objects.

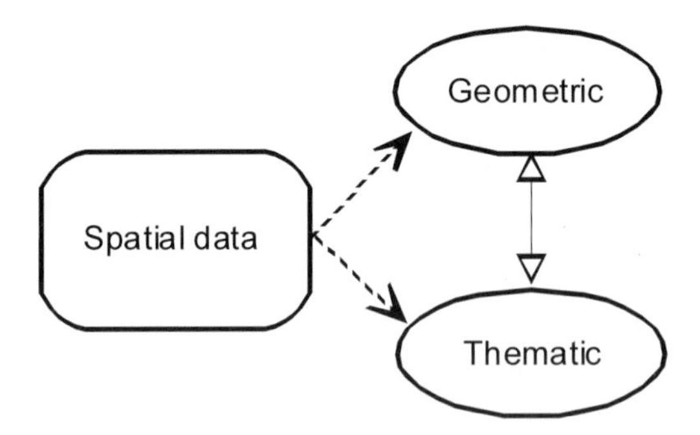

Fig. 4.1 Component of spatial data.

4.2 Spatial Data Modelling

Spatial data modelling is a process of describing real world spatial objects so that these objects as perceived by us can be represented in a form or notation which we understand and use. There are several techniques for perceiving the real world (Burrough and Frank, 1995). These techniques have different descriptive models for different levels of complexity of perception of the real world. If we would like to have these models represented and operational in a geo-information system, then they have to be mapped into data and processing models that can be handled by computers. Figure 4.2 illustrates a general view of the three stages of spatial data modelling that one may apply in information system development.

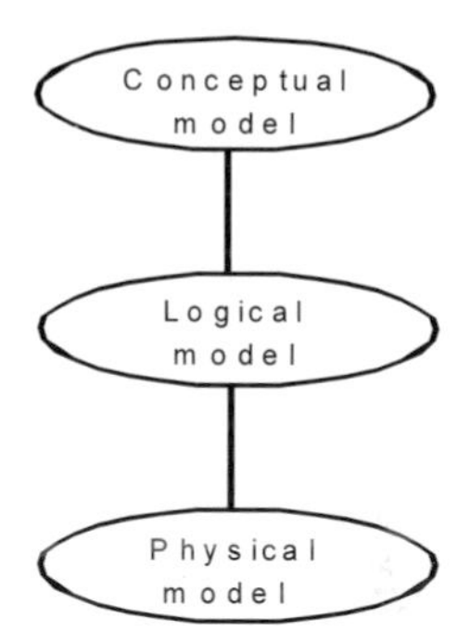

Fig. 4.2 A typical spatial data modelling steps.

A data model is a notation for describing data. It is a meta concept defining the content, structure, and meaning of data. The model also provides concepts to describe the structure and contents, for example, of a database, and the goal is similar to that of the data types (either basic data types or the Abstract Data Type (ADT)) used in programming languages to describe data within programs. Data models can be classified into the conceptual data model (or high level model), the logical data model (or implementation model), and the physical data model (or low-level model) as shown in Figure 4.2. Conceptual data models provide easy to perceive high-level concepts. They are used in the early stages of system development to communicate between end-users and system designer. Physical data models provide low-level concepts to describe how data are stored and accessed in the computer. The logical data model bridges the gap between the conceptual data model and the physical data model. It is sometime known as an implementation data model. It is used by database management system to implement reality in computerised databases. Figure 4.2 shows the steps in typical database design and also serves as a basic means to model terrain spatial objects.

4.3 Models and Their Importance for Geoinformation

Within the disciplines related to geoinformation science, the word 'model' has been used in two different ways. The first meaning is in the sense of a representation, or replica, of something regarded as real or genuine, like a globe in the classroom as a replica of the earth. The second meaning refers to something used to produce a number of replicas, and may be needed for the mass production of those replicas. The word 'model' in this sense may be comparable to the word mould, or form, and has the meaning of design, plan, or scheme. The quality of the mould directly influences the quality of the replica, so that more serious attention has to be paid to the design and construction of the mould than to the replica.

Regardless of the meanings of the word model, the process of producing a model is known unequivocally as modelling. It is necessary to state clearly what the model and modelling are actually meant for.

In the context of earth science, the end product we seek is a model in the sense of a replica of some portion of the planet earth, and is called a geo-spatial model. Since the term 'spatial model' covers a large territory over many disciplines (like the modelling of human anatomy in medicine, molecular structure in chemistry, or atomic structure in nuclear physics), we add the prefix geo to indicate the scope and purpose of this earth-related model.

For the information system to utilize the geo-spatial model, it must be constructed in a digital form, so that it can be maintained and exploited by a computer to perform certain tasks or operations that are:

1. less convenient in reality; for example, a distance can be obtained from a model instead of measuring from place A to B in reality, provided that places A and B are represented in the model
2. too expensive, too difficult, or practically impossible in reality; for example, a geologist may wish to see the continuous layer of sandstone lying fifteen metres under the earth's surface; removal of the upper soil to see this layer in reality is too expensive to contemplate.

The model in a digital form is in fact the database itself. Not only is a database a collection of data, it also contains relationships between data elements, and rules and operations to change the state of the data elements, regardless of how these components are stored. Components may be kept in one data set, or separately, at different places, depending on the system that manages and manipulates the model - the database management system (DBMS).

A model containing all aspects of reality is impossible, because of its complexity. Only some aspects can be included in the model at a manageable level. Hence, the quality of the model is judged only in terms of its purpose and how the model will be used. If the model permits the performance of the tasks or operations as required, and with acceptable results, the quality may be regarded as good. A model constructed for a single purpose may not be able to serve tasks or operations for different purposes, unless it is an integrated model designed and constructed for multi purposes.

A single-purpose model represents only a single view of the reality (Figure 4.3). Since an integrated model represents various views of the reality, such a model may be considered to be of higher value, since it contains more aspects of the reality and may serve more purposes.

4.4 Components of Geo-spatial Model

A model in the form of a database requires the categorization of aspects of reality into the components of the database managed and manipulated by a DBMS (Flavin 1981). The components of a geo-spatial model include the following:

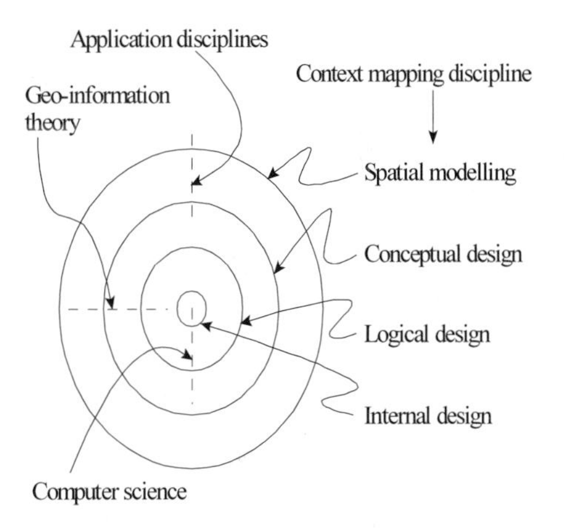

Fig. 4.3 Levels of geo-spatial modelling (after Molenaar 1994b).

1) Object types

Object types are classes of spatial entities in a geo-spatial model. In reality, they may be roads, rivers, cities, land use, and so forth.

2) Relationships

Spatial rrelationships are named associations between two or more spatial objects. For example, road A *passes through* city B. 'Passes through' defines a relationship between the road A and the city B, and may be written in a predicate form as 'Pass_through (road A, city B)' (Molenaar, 1994b).

3) Attributes, or descriptions

Attributes, or descriptions, are observed facts about a spatial object or relationship. An attribute or description is the smallest (non spatial) unit in the model, and has to be associated with an object type or relationship to be meaningful. An attribute or description cannot stand alone in the model. For example, the object type 'road' has the name 'A1', indicating that it is a highway passing by several cities.

4) Conventions

A convention results in a set of rules and constraints that govern the content, structure, integrity, and operational activity of the model. A convention applies to the entire model. An example, a convention stating that 'each feature class contains objects of only one geometric type' results in a rule preventing an area object from belonging to a line feature class (Molenaar, 1991).

5) Operations

A spatial operation is an action changing the state of the representation of a real world object being modelled, or deriving additional information from the current representation. Operations can be identified by events. Two types of operations can be distinguished: standard, and user-defined. Standard operations are provided for routine tasks. A user-defined operation is built by combining different types and sequences of standard operations. Standard operations include retrieve, add, delete, modify, union, intersect, difference, compare, and so forth. They can be applied to different components of the model.

4.5 Phases in Geo-spatial Modelling

Before continuing this review of the necessary fundamental concepts, the steps followed in geo-spatial modelling are defined.

Obtaining a geo-spatial model requires two main steps: the design phase, and the construction phase (see Figure 4.3). Once the model is in place, maintenance forms an additional phase. The design phase includes all the abstraction processes, ranging from the conceptual design, the logical design, to the physical design. The product of the conceptual design is referred to as a conceptual model, or data model (Peuquet, 1984; Maguire and Dangermond, 1991). It comprises a general scheme describing what should be included in the model.

The logical design sets out all the elements needed for the construction, without stating the actual size or type of each element of the model. This design results in a logical model, or data structure. The physical design phase specifies the actual size and type of each element of the model for the implementation of the geo-spatial model. For example, a 16-bit real number may be used to store the attribute 'width.' This phase yields an internal model, or file structure, to be used by the software engineer to establish the low level communication with the hardware at the bit and byte level. Figure 4.1 and Figure 4.3 graphically illustrate this process.

Molenaar (1994b) also suggests the involvement of different disciplines in geo-spatial modelling, as shown in Figure 4.3.

The five components of the geo-spatial model listed in the preceding section can be realized in these three different design phases: the object types and relations in the conceptual design phase; the attributes or descriptions of objects and relations in the logical design phase; the operations in the physical design phase.

The conventions must operate in every design phase. In the conceptual design phase, the conventions should state the allowable type of objects and relations between them to be included in the model. In the logical design phase, the conventions should state how the representation of one object is distinguished from another; an object should have a unique identifier. In the physical design phase, the conventions comprise a set of integrity and consistency rules for the operations that may change the state of the model; for example, the union of two areas sharing a common boundary has to yield only one area.

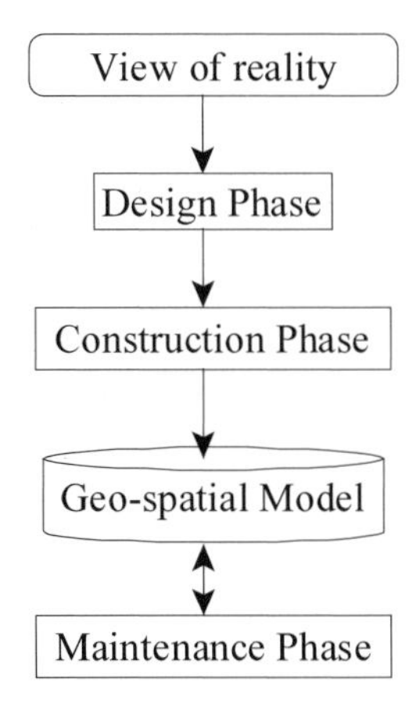

Fig. 4.4 Geo-spatial modeling.

The design of the model is followed by the design and implementation of the necessary functions and the user-interface to enable the construction and exploitation of the model. The result of this implementation is a geo-spatial information system (GIS). Having constructed the model, it must be kept valid to ensure that it remains in a state comparable with the reality, which is dynamic in nature. This is the maintenance phase. The basic maintenance operations of insertion, deletion, and modification can be applied to any component of the model, that is to say object types, relations, rules, attributes and operations. A GIS should also provide functionality to maintain the geo-spatial model.

4.6 Conceptual Design of a Geo-spatial Model

The design phase deals with the abstraction of reality into the representation scheme. This phase answers two basic questions: what aspects of reality (real world objects and the relationships between them) are to be modelled?; how should they be represented in the model?

A geo-spatial database represents a state of reality from a specific point of view or interest at an instant in time (if the temporal aspect itself is not the subject of the model). The reality consists of a set of various objects and the relationships between them which should be capable of representation as components of the model described in the preceding section. To be manageable, it is necessary to determine a limited number of aspects of the reality (objects together with the relationships between them) during the design phase which can be represented as the first and second components of the model.

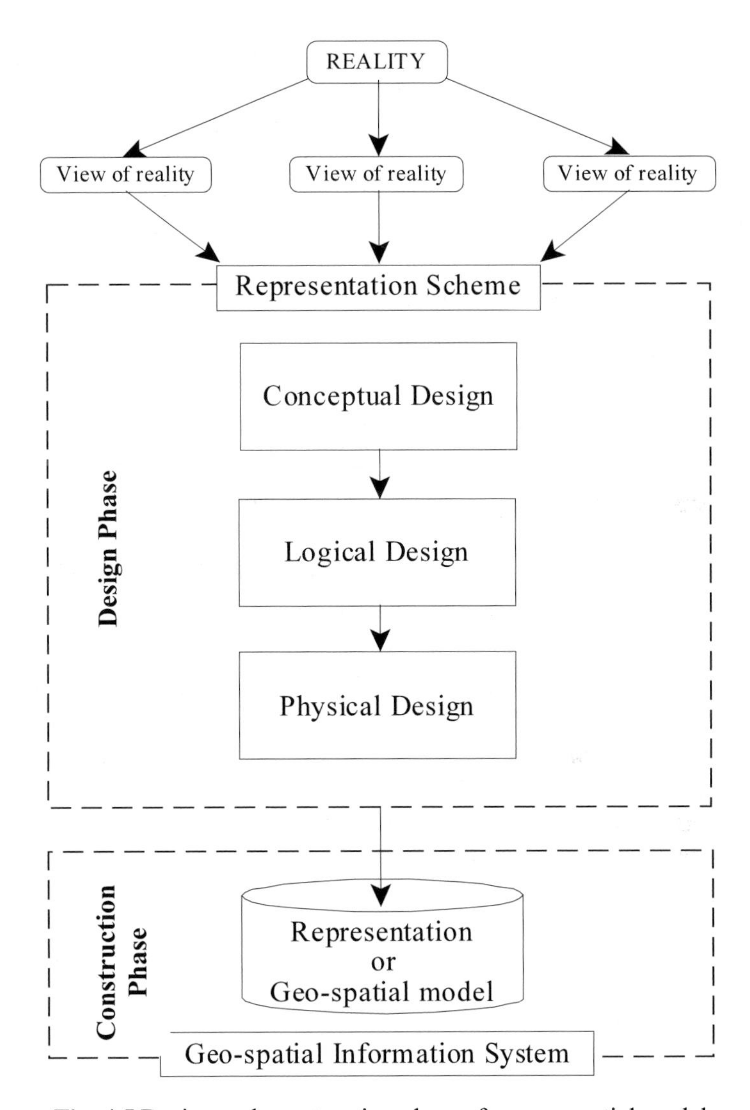

Fig. 4.5 Design and construction phases for a geospatial model

4.6.1 Definition of Space

Reality may be viewed as a space, that is to say, a collection of spatial objects and the relationships between them (Gatrell, 1991). Each spatial object occupies a subspace to define its own spatial extent, which may be defined by a set of spatial locations together with a set of interest properties characterizing those locations (Smith et al., 1987). Different sets of relations may define different types of space. Metric space, for example, is based on distance relationships; topological space is based on topological relationships.

For the mathematical description of space, we can rely on set theory, introduced by Cantor in 1880. Let O be a set of objects $\{o_1, o_2, o_3, ..., o_i\}$. R is a binary relation on O if $R \in O \times O$. If R is a relation on O, the relationship $(o_1, o_2) \in R$ can be denoted in prefix form by $R(o_1, o_2)$. A space S is then a collection, that is to say, a set of subsets, {[O], [R]}, denoted S = {[O], [R]}.

In reality, the space is an unbounded region consisting of numerous objects and relations. The space S (that is, a finite set) is only a view of reality in which the context is defined for describing the aspects of reality relevant to a particular discipline.

Having determined the collection [O] and [R], the question related to the aspects of reality to be modelled can then be answered. An example is only to include in the database the object types roads, rivers, buildings and land parcels and the relationships between buildings and land parcels, rivers and roads, roads and land parcels. In this sense, this database can be regarded as the space S.

To answer the second question, how to represent the objects of reality and the relationships between them, we have to consider some fundamental concepts of spatial modelling.

4.6.2 Abstraction of Space

There are two major abstractions of space, each of which passes on its characteristics to the spatial objects residing in that respective abstraction. The first conceptualizes space as tessellated into a contiguous set of smaller sub-spaces and is known as a field-based or tessellation-based definition. Each individual spatial object is composed of a set of subspaces (for example, a raster element in the raster-based GIS).

The second abstraction treats the space as empty and homogeneous, and consists of a collection of spatial objects. It is known as an object-based or feature-based definition (Ehlers et al., 1989).

Each sub-space of the field-based space is typically understood as, and associated with, regular shapes, like a square or a cube, usually found in the raster-based geo-spatial model. Irregular shapes are also used, such as in the triangular irregular network (TIN) that subdivides the space into a set of irregular triangular shapes, as frequently used for the representation of single value surfaces.

For object-based space, the best example is the vector-based geo-spatial model, where each object is composed of several vector elements in the form of geometric primitives (such as nodes, edges, faces, or bodies).

Field-based space and object-based space have different advantages and disadvantages. The field-based representation of space offers connectivity and continuity in all directions, thus providing the freedom to visit any location in space. An intuitive example from reality is travelling in free space in an aircraft. The pilot navigates by connecting the information in his vicinity such as landmarks, topography, or a city, to determine the travelling direction, but otherwise moves freely. This kind of approximation may be regarded as spatial interpolation.

Object-based representation, on the other hand, does not permit such freedom. The navigation in space is limited to a confined subspace defined by each spatial object. Connectivity and continuity are defined along with the existence of spatial objects. An example from reality would be travelling along a highway by car. The highway is comparable to a confined subspace of the global space. It restricts travel to a certain direction. The explicit destination is defined for each highway, so navigation in space is just a matter of selecting the right highway. No approximation for direction is necessary in this case.

The abstraction of space is typically decided during the conceptual design phase, which is usually driven by the type of spatial operations. In this book, we present an attempt to integrate the field-based and object-based abstractions into a hybrid abstraction to allow confined and unconfined navigations in a geo-spatial model, thereby facilitating a wider range of spatial operations.

4.6.3 Abstraction of Real World Object

In the present context, the earth is the subject under consideration. It is important to bear this in mind, since some aspects of the earth have to be taken into account and included into the model. In geoinformation science, any real world object may be described geometrically and thematically (see Figure 4.4 and Molenaar, 1989, Maguire et al., 1991, Gatrell, 1991). The terms metric and semantic have also been used (Makarovic, 1984). In this book, the representation of a real world object is referred to as a feature where the terms spatial entity and geo-object may be found elsewhere (Peuquet, 1988; Laurini and Thompson, 1992; Raper, 1989). A real world object that has to be described, or related to a location in reality, is referred to as a spatial object.

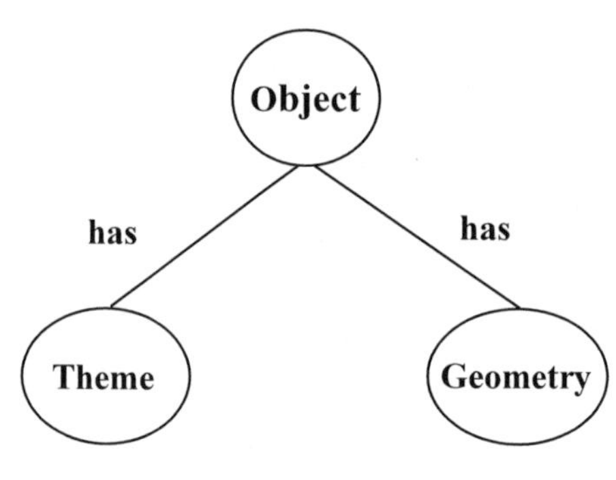

Fig. 4.6 A general abstraction of the real world object (at an instant of time)

Figure 4.6 can be regarded as a general representation scheme for any spatial object. The thematic and geometric aspects may be separately modelled and considered as general component types of the model. They have, however, to be brought together at some stage. The geometric aspects are the spatial characteristics of the object such as shape, size and location. The thematic aspects are the non spatial characteristics of the object related to its state, functionality, or utility in reality.

Figure 4.6 provides an extreme level of abstraction about the aspects of the reality we want to deal with. It can only be used as a general framework for the overall modelling process. This abstraction must be further elaborated to achieve a more specific design.

Note, however, that Figure 4.6 is limited to the representation at a certain instant. Object states that change over time have not been considered. Otherwise, the dynamic aspects of the objects would have to be included as additional components of the model. The term spatio-temporal is used to indicate such a kind of model; it lies, however, outside the scope of this book (Langran, 1992; Tansel et al., 1993 for more details on this subject).

4.6.3.1 Geometric Component

Two important aspects of a real world object, location and shape, need to be included into the model which allows further derivation of the size of the object. However, they can only be correctly described if the dimension of space is taken into account. In mathematics, the dimensions of Euclidean space are expressed through the number of referential axes, which are linearly independent from each other. The distances from the origin along each axis are arranged into the *ordered-n-tuples* notation $(a_1, a_2, \ldots, a_n)$, which is a sequence of n real numbers used to represent a coordinate tuple in *nD-space* (denoted by R^n, see also Anton, 1987).

In geoinformation science, the term 'dimension' has been used to denote various meanings. With respect to boundary representation, the dimension may indicate the data type being used to represent the object, such as point (0D), line (1D), area or surface (2D) or body (3D) (Frank and Kuhn, 1986). Each object occupies a subspace and has its own dimensionality, which may be regarded as the internal dimension. The external dimension (R^n) is then the dimension of the space embedding the object. The term 'dimension' also frequently indicates both the internal and external dimensions. For example, 2D may mean objects in R^2, 2.5D means 2D objects in R^3, 3D means 3D objects in R^3, 3.5D means 3D objects in R^4, and so on. In this book, these kinds of notations are used, dependent on the context of each part. Egenhofer and Herring (1990) also discuss the dimensionality of space and objects.

Having defined the dimension, an object's location and shape can be described. Location is defined by a set of coordinate tuples, while the description of shape can be given in different ways, for example, by a mathematical function, a verbal description, or a skeleton with radius functions (Blum, 1967; Pilouk, 1992). CAD is a discipline focusing on the 3D modelling of geometric aspects where several approaches have been used:

- **Primitive Instancing (PI):** describes an object by a set of parameters together with a shape function; for example, a rectangle can be described by its width, height and a rectangular shape function.
- **Sweep Representation:** applies to an object of regular shape; for example, a cylinder is the result of sweeping a circle along a straight line.
- **Boundary Representation (BR):** describes an object through its boundary elements, that is to say, the vertices, edges and faces for a 3D object.
- **Constructive Solid Geometry (CSG):** hierarchically decomposes an object into a set of components with simpler geometry. The node of each hierarchy may contain the set operator needed to combine together the components in a lower level of the geometric hierarchy. Translation and rotation parameters may also be attached to this node. For example, a solid cube with a cylindrical hole can be decomposed into a solid cube and a solid cylinder under the set operator $\cap$ (intersection) and a translation parameter to align the cylinder into the middle of the cube.

- **Spatial-Partition Representation:** also decomposes an object similar to CSG, but to the more primitive level known as cell. Only the set operator ∪ (union) is allowed to combine cells to reconstruct the object. This means that no intersection between two cells is possible. This distinguishes the spatial-partition representation from CSG. One criterion for this kind of decomposition is that the adjacent cells must share common boundary elements, such as vertices, edges, or faces. Different decomposition schemes can be used:
  - **cell decomposition:** decomposes an object into various types of primitives with shapes which need not be regular; for example, a simple house may be decomposed into a cube and a prism.
  - **spatial-occupancy enumeration:** decomposes an object into a set of regular cells with fixed shape and size; for example, regular-grid, pixel and voxel.
  - **irregular tessellation**: decomposes an object into one type of primitive with, different shapes and sizes; for example, triangular irregular network, tetrahedral network (TEN).
  - **hierarchical regular subdivision:** subdivides a space into homogeneous zones using only one type of primitive which varies in size; for example, rectangle (quadtree), cube (octree).
  - **binary space-partitioning (BSP):** subdivides an object into pairs of planes with arbitrary orientation and position.

In describing an integrated model, the boundary representation with irregular tessellation is used to geometrically describe an object. More details about geometric modelling can be found in Requichar (1980), Mäntylä (1988), Samet (1990), Foley et al. (1992), Bric (1993), and Cambray (1993).

4.6.3.2 Thematic Component

Apart from geometry, objects are given a referential identifier and descriptions and may be organized into a group, or theme, to differentiate them and make reference to them more convenient. Objects having the same characteristics may be grouped together, becoming more easily distinguished from objects with other characteristics. Nevertheless, the criteria for judging whether an object belongs to a particular group are based on a specific viewpoint. Using different criteria, an object can be classified into a different group. The process of classifying objects into groups is known as thematic modelling. The term single-theme is used when the geometric description of an object is related to only one theme (Molenaar, 1989), and multi-theme if the geometric description of an object relates to more than one theme (Kufoniyi, 1995).

In this monograph, single-theme and multi-theme express the homogeneous and heterogeneous properties of a spatial object.

Since thematic modelling is context dependent with respect to a particular application domain, the work discussed in this book made no attempt to achieve the modelling of a thematic component capable of accommodating a wide range of applications.

4.6.4 Object and Spatial Extent

There are two kinds of spatial objects that can be distinguished on the basis of knowledge about their spatial extent. The first is of the type determinate spatial extent. Objects of this type are referred to as determinate spatial objects which have discernible boundaries, and are typified by houses, roads, rivers, land-parcels that can easily be sensed. Spatial objects of the second category have indiscernible boundaries which are difficult to sense. These objects are of the type indeterminate spatial extent, for example colloid in water, plume of smoke, temperature distribution, soil type, etc., and are referred to as indeterminate spatial objects. The boundaries of determinate spatial objects can be sampled and directly represented in the database. However, this is not the case for indeterminate spatial objects - their boundaries cannot be directly sampled and must be derived by means of classification, or computation, using specific property values of the surrounding neighbours (for example, by interpolation or extrapolation). Therefore, the representation of indeterminate spatial objects in the database can only be indirect.

4.6.5 Spatial Relations

Spatial relations are a key issue in the design of a spatial model. Many extensive reviews and discussions can be found; Frank and Kuhn (1986), Pullar (1988), Pullar and Egenhofer (1988), Egenhofer (1989), Egenhofer et al (1989), Kainz (1989), Egenhofer (1990), Kainz (1990), Egenhofer and Franzosa (1991), Pigot (1991), Pigot (1992). This section provides only a brief review of some of the important basic concepts in spatial relations.

A set theoretical definition of a relation has been given in section 4.6.1. Recall that R is a relation on a set O of objects. In general, R can be further distinguished by its different basic properties that depend on the relationships between its member elements (see also Willard, 1970; Stanat and McAllister, 1977; Pullar and Egenhofer, 1988).

- R is reflexive, if each element can be compared with itself (if and only if $(o_i, o_i) \in R$), for example 'point A' is equal to itself.
- R is symmetric, if and only if $R(o_1, o_2)$ implies $R(o_2, o_1)$. For example 'area A' is adjacent to 'area B' implies that 'area B' is adjacent to 'area A.'
- R is antisymmetric, if and only if $R(o_1, o_2)$ and $R(o_2, o_1)$ implies $o_1 = o_2$ for all $o_1, o_2 \in O$, for example if $a \leq b$ and $b \leq a$, then $a = b$.
- R is transitive, if and only if $R(o_1, o_2)$ and $R(o_2, o_3)$ implies $R(o_1, o_3)$ for all $o_1, o_2, o_3 \in O$, for example area A < area B and area B < area C then area A < area C.

For example, given a set of real number N, $<$ is a transitive relation on N, $\leq$ is a reflexive, antisymmetric, transitive relation on N, and $\neq$ is a symmetric relation on N.

It is necessary at this stage to consider the definition of functions in mathematics used later.

Given two sets A and B, a function (or map) f from A to B, denoted $f: A \rightarrow B$, is a subset of the Cartesian product A x B with the following properties:

a) For each $a \in A$, there is some $b \in B$ such that $(a, b) \in f$.
b) If $(a, b) \in f$ and $(a, c) \in f$, then $b = c$.

Each $a \in A$ must be in relationship with exactly one $b \in B$ and the relationship $(a, b) \in f$ is normally written in a prefix form as $b = f(a)$.

Comparing with relation R, every function on A is a relation R on A. However, not all relations on A are functions.

Three classes of spatial relations, namely metric, order and topology, have been distinguished, based on the type of function or relation associated with a set of objects (Egenhofer, 1989).

Metric

Metric relations are built around the notion of distance function. Its mathematical description is as follows (see also Willard, 1970):

Given a set M with x, y, z $\in$ M and a set of real number N. A metric relation d is a function d : M x M $\rightarrow$ N with the following conditions:

a) $d(x, y) \geq 0$, Distance from x to y is more than or equal to zero.
b) $d(x, x) = 0$; $d(x, y) = 0$ implies $x = y$, Distance from x to itself equal to zero. Distance from x to y equal to zero implies that x is equal to y.
c) $d(x, y) = d(y, x)$, Distance from x to y equal to distance from y to x
d) $d(x, y) + d(y, z) \geq d(x, z)$ (triangular inequality). Distance from x to y plus distance from y to z is more than or equal to distance from x to z.

A **metric space** is an ordered pair (M, d) consisting of a set M together with a function d: M x M $\rightarrow$ N satisfying the above four conditions. The function d is also called the metric on M. Functions d : M x M $\rightarrow$ N are called distance functions. A metric space is the Euclidean n-space, denoted R^n, if the distance function is the Euclidean distance below:

$$d((x_1, \ldots, x_n), (y_1, \ldots, y_n)) = \sqrt{\sum_{i=1}^{n} (x_i - y_i)^2}$$

The number ‘n’ defines the number of distance components between x and y (each one computed along an independent vector) and denotes the dimensionality of Euclidean space.

The distance functions available in metric spaces are used to develop the notion of continuity crucial for the development of topology.

Order

Order defines a comparative type of relationship between the objects based on a preference. Two kinds of order relations can be distinguished: strict order and partial order. Strict order is a relation $<$, which is transitive. This kind of relationship may be represented as a tree-like structure. Partial order is a relation $\leq$, which is reflexive, antisymmetric, and transitive, and may be viewed as a network structure. Every order relation has a converse relationship, for example $a < b$ conversely implies $b > a$. A formal study about the use of order for spatial relationships has been reported by Kainz (1989), Kainz et al., (1993). Algorithms and data structures for order operations have also been presented in Kainz (1990).

Topology

Since the eighteenth-century, topology has developed as a discipline of mathematics. The definition of topology, as the study of the properties of figures remaining invariant under topological transformation, was given by Augustus Möbius (Devlin, 1994). The explanation of topology in this section, however, follows the general (point set) topology founded by Hausdorff in 1914 (see Willard, 1970). Point set topology was developed from metric (distance) which is easier to understand (Mäntylä, 1988). The purpose of introducing topology is to be able to define any continuous function without mentioning distance (Willard, 1970; Armstrong, 1983; Pullar and Egenhofer, 1988), thus adding the concept of 'neighbourhood' to location, distance, and direction (Kainz, 1989). The expression of spatial relationships in the form of topology is more appropriate for handling by current computer technology, which bases arithmetic computation on a finite numbering system, and so cannot be used to completely represent continuity based on Euclidean distance (Franklin, 1984; Frank and Kuhn, 1986). For example, the state of a point lying inside a polygon might be changed after rotation or scaling, because of rounding errors.

The expression of continuous function is accomplished by introducing the concept of a point-set in metric space that is an open set (a set that does not include its boundary; Pigot, 1991). Any open set has a continuity property (consult Willard (1970) for proof). A point-set P is an open set if every point $x \in P$ is surrounded by an ε-sphere of radius $\varepsilon > 0$ such that distance between x to any point $y \in P$ is always less than ε. An example of ε-sphere about a point c of a set of real number is an open interval $(c-\varepsilon, c+\varepsilon)$.

A mathematical definition of topology is as follows:

A topology on a set X is a collection T of subsets of X, called the open sets, satisfying:

a) Any union of elements of T belongs to T,
b) Any finite intersection of elements of T belongs to T,
c) $\varnothing$ and X belong to T.

A topological space is denoted by (X, T). Given two topological spaces, A and B, $f : A \rightarrow B$ is a continuous function if it preserves the neighbourhood relations between mapped points. This mapping is also called continuous mapping, or homeomorphism (Alexandroff, 1961; Pigot, 1991). The topological transformation is commonly known as rubber sheeting, in which translation, rotation and scaling are included (Pullar and Egenhofer, 1988). Examples of homeomorphic mappings are transformations to correct distortion resulting from paper or film shrinkage in cartographic or photogrammetric digitizing.

Some properties of topology have been expressed as follows:

- Topology is defined as the set of properties which are invariant under homeomorphisms (Alexandroff, 1961) - one-to-one, continuous, and onto transformation (Pigot, 1991).
- Topological relationships are invariant under topological transformations such as translation, scaling, and rotation (Egenhofer, 1989).

The transformation that includes translation, scaling, and rotation is known in photogrammetry as geometric transformation. It defines changes in shape, size and location of an object.

Topology describes the relationship between an object and its neighbours. Topological relations can be defined through the three components of an object, that is, the interior, boundary, and exterior (Vaidyanathaswamy 1960; Pullar, 1988). An elementary set operation, an intersection, is used as a mechanism to determine each type of relation. For example, if the intersection between a boundary set of object A and a boundary set of object B yields a non empty set, the relationship between the object A and the object B may be defined as 'touch.' If in addition the intersection between the interior set of the object A and the interior set of the object B also yields a non empty set, the relationship between these two objects may be defined as 'overlap.' Figure 4.7 shows some examples of topological relationships between two objects.

Intersection of the three components of two objects can be organized into a 3x3 matrix. This gives a 9-digit logic state, called a 9-intersection, which can be interpreted as relation codes (see Bric 1993). The 9-intersection gives in total 512 possible relationships, from which a set of relevant relationships can be found by a process described in Pullar and Egenhofer (1988).

The dimensionality of topological space has frequently been mentioned in the literature. Since topological space is derived from metric space, it also inherits the dimensionality defined in metric. 2D topology would mean that the topological relations are only valid for 2D metric space and certainly 3D topology would only be valid for 3D metric space. For example, a triangle may have at most three neighbours in 2D space, but there can be many more neighbours in 3D space. An important limitation is that there is no continuous mapping from the higher dimension to the lower dimension. This endeavour will therefore result in loss of information. 2D topology has been intensively studied by Egenhofer (1989). Pigot (1991), Bric (1993), Rikkers (1993) have studied 3D topology, while Pigot (1994) have studied 4D topology.

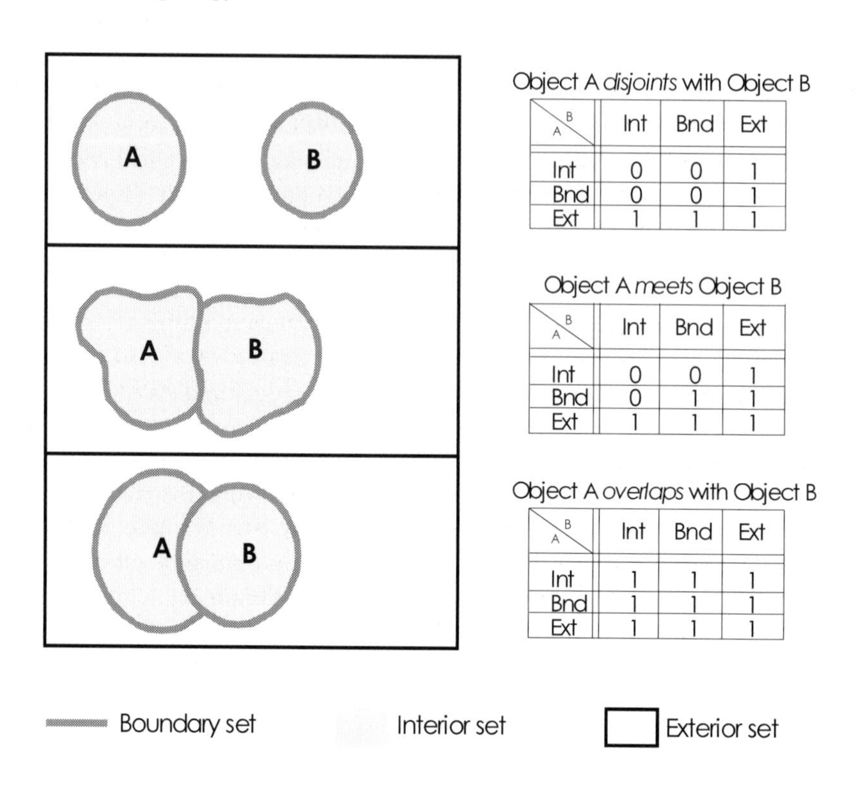

Fig. 4.7 Example of spatial topological relationships.

4.6.6 Application of Spatial Relations

The three types of spatial relations have been used in GISs with little realization by users concerning their categories. This section identifies some of their usage with respect to type of relations.

Spatial Indexing

A spatial database often contains a large volume of data, so a lot of time is required for the data retrieval process, particularly for non sequential data access, for example during a query operation. This process can be speeded up by heuristically limiting the search space in the database. We must know roughly the location in the file that contains the data elements or records. Storing a bounding rectangle for a set of data elements can be used as a method of giving a rough spatial index to a subset of the database. By using the metric relation to compare the coordinate set of the data to be retrieved with the coordinates of two opposite corners of the bounding rectangle, faster data retrieval can be achieved, because the search is limited within a particular rectangle at the end. Bounding rectangles of many subsets of a database may be further organized using a tree structure, further speeding up spatial access. An example is the R-tree (Guttman, 1984; Samet, 1990) which exploits the order relation to organize the 2D data and allows the search to proceed from coarse to fine. Navigation in the tree structure also helps avoid metric computation requiring a long access time, so it dramatically speeds up the process. Other examples of spatial indexing using an order relation are quadtree and octree (Samet, 1990). Topology can also be used for spatial indexing by storing the links (for example, pointers) between data elements in the database directly. However, the storage of such information is redundant to the storage of coordinate sets based on metric relations. Problems of consistency between metric relation versus topology or order relation arise. The consistency rules must be defined and enforced to eliminate conflicts for any database operations that may change the status of the database, for example to insert, delete or modify a data element in the database. Examples defining and applying consistency rules for spatial database can be found in Kufoniyi (1995).

Spatial Analysis

Two kinds of spatial analyses may be distinguished: query-based, making preferential use of topological, and order relations, and computation-based, relying heavily on metric and order relations. Spatial relationships like 'touch' or 'disjoint' can be expressed in terms of metric relations. Peuquet (1986) has defined a relationship 'touch' by a distance equal to zero and never less than zero at a single location. The relationship 'disjoint' may be defined so that 'the distance from any point of object A to any point of B is greater than zero' (Egenhofer, 1989). A distance relationship can be expressed in different forms, for example direction and proximity, which are commonly used in spatial modelling. Based on some referential axes, distances can be used for georeferencing in the form of a coordinate tuple and can be transformed into directions in terms of angularity. Discrete directions, for example north, east, south, west, can be used to express spatial relationships (Alia and Williams, 1994). Direction may further define an orientation, for example 'from-to', useful for path finding in a network structure. Proximity can be used to represent spatial relationships like 'near', 'far' or 'within the distance of', for example in the form of a buffer zone, or distance tolerance. The distance tolerance always needs to be defined for metric operation, because of the finite state of the computer, which results in rounding-off errors. For example, intersection or touching between two straight lines may be encountered and then recorded as topological relationships in the database, if the shortest distance between the two lines is less than a predefined distance tolerance. Combining metric and order relations is typically used for the elimination of short straight line segments. The length of a straight line, that is the distance between its two nodes, is computed and then the order relation $<$ is used to compare it with the predefined distance tolerance to decide on the deletion of this line.

Operations based on metric relationships (known as computational geometry) are time consuming if the data elements have not been organized in an appropriate structure, lacking order or topology. However, most raster operations, which are mostly simple and fast, are based on metric relationships, because the data elements have already been organized into a strict spatial order. Many relationships can be interpreted as order, for example, in front/behind, larger/smaller, greater than/less than, under/over, higher/lower, equal. For instance, the operator $\leq$ is spatially interpreted as 'is contained in' and can be used to answer a question like 'what land parcels are inside the zone A?'

The use of topology has gained significant attention in GIS. It is used to help the navigation between data elements in the database without using sophisticated computational geometry. This is usually done by explicitly storing the topology in the database which may be initiated manually by human knowledge, or analytically derived by computational geometry (up to some accuracy). Topological relations are translated into different types of pointers from one data element to another and can therefore speed up the searching operation, because the search space is dramatically limited. For example, the incident relation 'meet' defined for polygons A and B yields an arc C that is the result of the intersection between the boundaries of A and B. The inverse relationship may be defined from the arc C to the two polygons A and B as 'left' and 'right' of the arc respectively.

$$\begin{aligned}&\text{meet}(A, B) = A \cap B = C;\\&\text{left}(C) = A;\\&\text{right}(C) = B\end{aligned}$$

The 'left' and 'right' pointers are then stored in the database directly. Navigating in the database from A to B can be achieved by starting from A and then searching for B exclusively via C.

4.6.7 Representation of Spatial Objects and Relationships

Models of the earth and geo-spatial objects may be physically created; examples are a metal or plastic globe used for teaching geography, a plaster magnet of an urban quarter. Creating a model of the earth in the computing environment requires different modelling tools. For the vector type of geoinformation, objects have to be represented in an appropriate form convenient for storage, analysis, or graphical display. Two levels of representation of spatial objects are widely used. The first has its roots in the concept of cell complex and simplicial complex (Frank and Kuhn 1986). The other is in the concept of graph theory. The real world objects are mapped to different types of elements of the two representations, depending on the level of abstraction, as explained in the following sections.

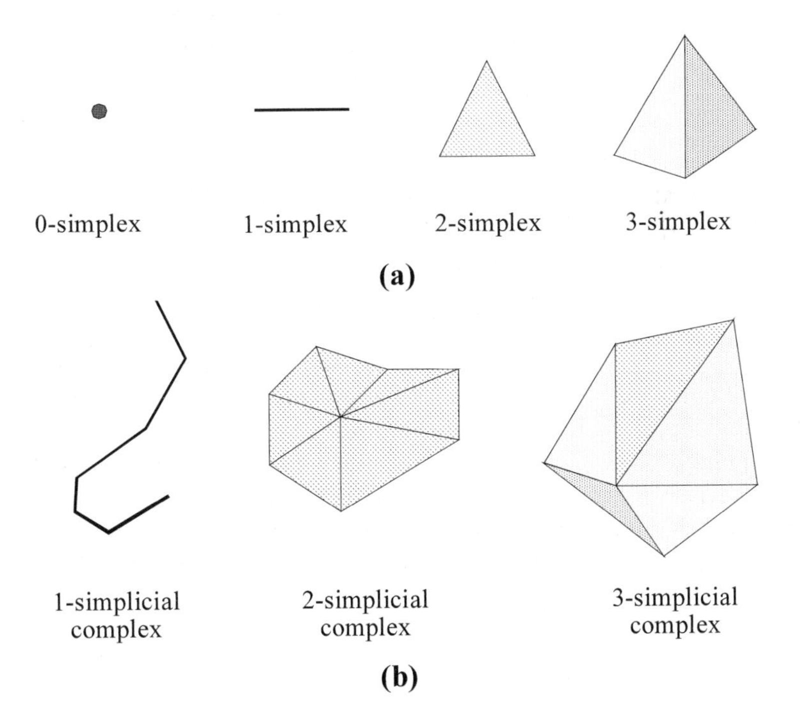

Fig. 4.8 Examples of (a) simplices and (b) simplicial complexes. The hull of each simplicial complex is a cell-complex.

Definition Level: Cell Complex and Simplicial Complex

Based on the terminology used in Moise (1977) and Giblin (1977), Frank and Kuhn (1986) described two types of elements, namely complex and simplex, that can be used for the representation of real world objects.

A formalization of this concept has been provided by Egenhofer (1989). A complex constitutes a description of an object as a whole. Different types of complexes are defined by the (internal) spatial dimensions of objects, that is 0-cell (point-object), 1-cell (line-object), 2-cell (area-object), 3-cell (volume-object), and so on. However, for any spatial dimension, there is a simplest geometric figure that can represent an object. This type of geometric figure is called a simplex. For example, every point (node) which is a geometry of dimension zero, is a 0-simplex. For spatial dimension one, a straight line segment is the simplest geometry, so it is a 1-simplex. Likewise, a triangle is a 2-simplex and a tetrahedron a 3-simplex in two and three dimensions respectively. In general, an n-simplex is the simplest geometry of dimension 'n.' A mathematical definition of simplex is as follows:

Any simplex of dimension n, called n-simplex, is bounded by (n+1) geometrically independent simplices of dimension (n-1) (Egenhofer et al., 1989).

For example, a tetrahedron is a simplex of dimension 3, that is, a 3-simplex. It is bounded by (3+1) = 4 triangles that are simplices of dimension 2. These four triangles are not components of each other and do not coincide; that is, they are geometrically independent. A triangle, that is, a 2-simplex, is bounded by (2+1) = 3 edges. Likewise, an edge is a 1-simplex bounded by two nodes that are 0-simplices. Figure 4.8 shows some examples of simplices and simplicial complexes.

If a complex object is composed of a contiguous and finite set of non overlapping simplices, it is called a simplicial complex. For example, a line object composed of a chain of straight line segments is a 1-simplicial complex. A polygon composed of a set of triangles is a 2-simplicial complex. The hull of a simplicial complex is a cell-complex. The concept of a simplicial complex is crucial for our design of an integrated data model.

Description Level: Graph

A simplicial complex is a representation of geo-spatial objects at a definition level which still requires further elaboration. Understanding is usually facilitated by making an idea perceptible. A graph representation provides such a possibility. The elements of a simplicial complex can be mapped to elements of a graph which allows the idea to be visualized.

The origin of graph theory is attributed to Leonhard Euler, the Swiss mathematician, with his publication in 1736 of an answer to the question known as the 'Bridges of Königsberg' (Finkbeiner and Lindstrom, 1987; Devlin, 1994). It is interesting to note that the first use of a graph was to solve a problem of a spatial nature. Graph theory has been applied in many disciplines, for example, electrical engineering, artificial intelligence, information modelling. In geo-spatial modelling, graph theory has been applied to represent the topological structure of geographic databases, such as in DIME (Corbett, 1979; Marble et al., 1984), TIGER, ARC/INFO (1991), FDS (Molenaar, 1989). It is also applied in many parts of this research. Corbett (1979) has extensively applied concepts of graphs to express the topological relationships between the cell complexes that are elements of a cartographic model. Graphs have also been used as a tool to assist spatial network analysis, such as finding the optimum path, or the shortest path. This section reviews some fundamental concepts of the graph.

Different graph elements have been provided for the representation of objects in the same way as elements provided by the concept of simplicial complex. Traditional elements of a graph are node, edge and face. However, these are limited to 2D representation. Additional elements of a graph are needed for the representation of objects that have a greater spatial extent than 2D; for example a 'body' element for a 3D object. In order to compare the simplicial complexes and graphs, we first look at the definition of the graph using mathematics.

Definition of a Graph

A *graph* G is defined by an *incident* relation between two disjoint sets N and E, where N is a non empty set of i nodes ($N = \{n_1, n_2, n_3, \ldots, n_i\}$) and E is a set of j edges ($E = \{e_1, e_2, e_3, \ldots, e_j\}$). If E is an empty set, a graph is called an *empty,* or *null graph.* If both N and E are finite sets, a graph is called a *finite graph.*

An edge *e* is further defined by a set of two nodes $\{n_1, n_2\}$ with *an adjacent* relationship. These two nodes can either be the same, or different. If an edge is incident with two nodes that are identical, this graph is called a *loop*. A graph that contains no loop is called a *simple graph*. If a graph contains a pair of nodes that are incident with more than one edge, the graph is called a *multigraph*. Figure 4.9 show some example of such graphs.

The adjacent relation can also be defined for a set of edges; the edges are said to be adjacent if all of them are incident with the same node. Here, the *degree (or valence)* of a node *n* of a graph can be defined by counting the number of edges that are incident at *n*.

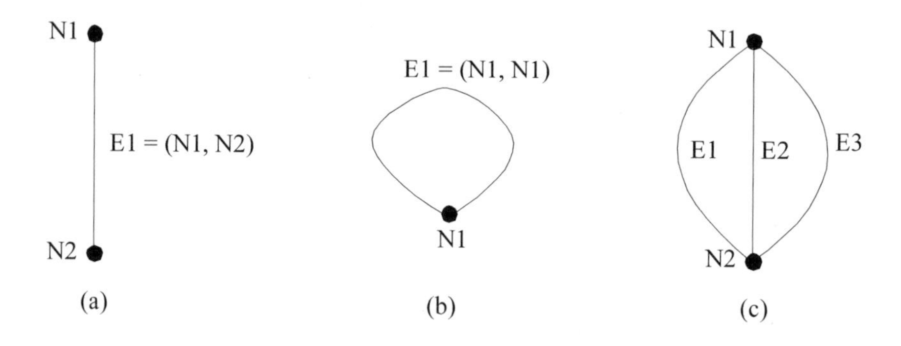

Fig. 4.9 (a) a graph, (b) a loop and (c) a multigraph

Types of Graphs

If every node of a graph is of the same degree, the graph is a regular graph. An edge, a triangle, a tetrahedron and a cube, are examples of graphs of this kind with nodes of degree 1, 2 and 3, respectively. A graph with m nodes is a complete graph, denoted by K_m, if each pair of distinct nodes is joined by one edge.

The degree of a regular graph is comparable to the dimension number of a simplex, as defined in the preceding section. A complete graph K_m is equivalent to a *(m-1)-simplex*. This comparison is shown graphically in Figure 4.10.

Graphics			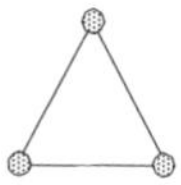	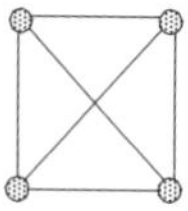	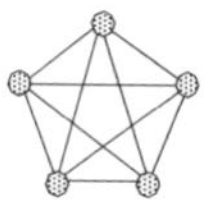	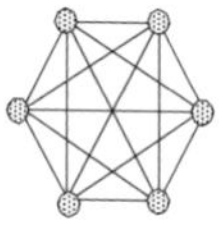
Simplex	0	1	2	3	4	5
Complete graph	K_1	K_2	K_3	K_4	K_5	K_6
Name	Node	Edge	Triangle	Tetrahedron	...	...

Fig. 4.10 Complete graphs and simplices

If a set of nodes of a graph G can be divided into two non empty and disjoint subsets M and N of m and n nodes respectively, such that an edge of G connects with a node of M and a node of N, this graph is called a *bipartite graph*. If each node of M has a distinct set of edges connecting every node of N and vice versa, then G is a *complete bipartite graph* and is denoted $K_{m,n}$ (see Figure 4.11).

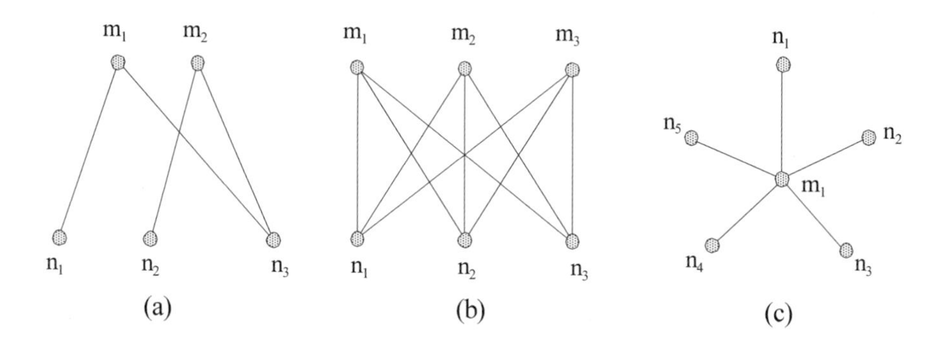

Fig. 4.11 *Bipartite* graphs (b) $K_{3,3}$ (c) $K_{1,5}$ or star graph. Only (b) and (c) are complete *bipartite* graphs

A graph can also contain subgraphs. If G_1 and G_2 are graphs with $\{N_1, E_1\}$ and $\{N_2, E_2\}$ incident relationships of nodes and edges respectively, G_2 is a subgraph of G_1 if and only if all nodes of N_2 are nodes of N_1 and all edges of E_2 are edges of E_1.

Similarity of Graphs

The similarity of graphs can be expressed more formally in terms of isomorphism and homeomorphism. Two graphs $G = \{N, E\}$ and $G' = \{N', E'\}$ are *isomorphic* if and only if N and N' are one-to-one correspondent and E and E' are also one-to-one correspondent.

To define the homeomorphism of two graphs, the concepts of subdivision and contraction have to be considered. Let a graph $G = \{N, E\}$ and an edge $e = \{n_1, n_2\} \in E$, then a simple subdivision G' of a graph G can be obtained by inserting a new node m of degree two on e, therefore between n_1 and n_2, and replacing the edge e with two new edges incident to two pairs of nodes $\{n_1, m\}$ and $\{m, n_2\}$. If G' results from a sequence of one or more simple subdivisions of G, G' is called a subdivision of G. Conversely, if a graph $G = \{N, E\}$, and two edges $e_1 = \{n_1, n_2\}$, $e_2 = \{n_2, n_3\} \in E$, then a contraction of G can be obtained by deleting the node n_2 (of degree two) and consequently replacing e_1 and e_2 by the new edge $e = \{n_1, n_3\}$. It is possible to perform this kind of contraction for every node of degree two of a graph. Subdivisions and contractions are inversions of each other.

The two graphs G and G' are said to be homeomorphic if they are isomorphic, or if G' is either a subdivision or a contraction of G. The concept of homeomorphism is useful for the generalization of some aspects of graphs. Some examples of isomorphic and homeomorphic graphs are shown in Figure 4.12.

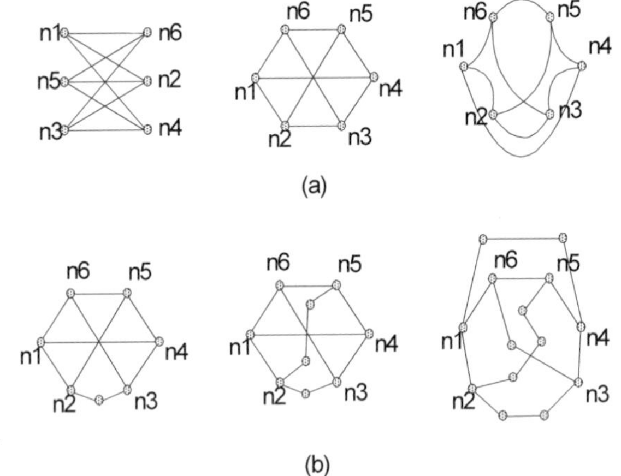

Fig. 4.12 (a) Graphs that are isomorphic; (b) Graphs that are homeomorphic to (a)

An intuitive approach can be used to verify the isomorphism of graphs: if two graphs G_1 and G_2 have different numbers of nodes or edges, or if none of the nodes of G_1 graph are of the same degree as any node of G_2, then G_1 and G_2 are not isomorphic.

Connectivity of Graphs

One aspect of graphs frequently applied to spatial modelling is connectedness. A graph can be connected or disconnected. For a connected graph, every edge belonging to a walk W that visits all the nodes and edges of the graph also belongs to the graph, while the same kind of walk for a disconnected graph cannot be defined without introducing an extra edge that is not an edge of the graph. It can also be said that a connected graph has only one component, while a disconnected graph has two or more disjoint components. For example, a cadastral map may be regarded as a set of graphs used to represent different objects, such as houses and land parcels. If some individual houses situated as islands inside land parcels are contained in the map, it can be thought of as a disconnected graph. In this monograph, the number of disjoint components is expressed as *the degree of isolation*. This aspect is also important in the generalization of Euler characteristics of graphs.

Planarity and Non Planarity of a Graph

The planarity of the graph is also commonly applied for checking the internal geometric consistency in a 2D-based spatial model (see Laurini and Thompson, 1993); for example, an edge must have two different nodes stored in the same database, and a triangle must have three edges. By definition, a graph G is planar if and only if it is isomorphic to a *plane graph* that can be drawn on a plane with no crossing edges. Another definition states that a graph is planar if and only if it can be embedded on the surface of a sphere (Wilson, 1985). The latter definition allows a planar graph to be embedded in a 3D Euclidean space so that the edges that seem to cross each other when embedding onto a plane can be placed separately around the surface of a sphere.

The algebraic approach to verifying whether a graph is planar is by applying the Euler equality.

Theorem (Euler): Let G be a connected, plane graph with n nodes and e edges. Then G should satisfy the equation

$$n - e + f = 2$$

where f is the number of non overlapping regions separated by G. Note that each face of a planar graph is bounded by a simple circuit if the face is finite. The right side of the equation is called the *Euler characteristic* of a planar graph.

It is important to note that the outer (infinite) region has to be counted as a face; otherwise, the formula becomes

$$n - e + f = 1$$

Also, to cover a disconnected graph with i components, the Euler equality has been generalized as follows (see also Wilson 1985):

Theorem (generalized form of Euler's theorem): Let G be a planar graph having i components, n nodes, e edges, and f faces. Then G should satisfy

$$n - e + f = i + 1$$

The outer region should be counted only once. If it is not counted, then the formula becomes

$$n - e + f = i.$$

This generalization is equal to the summation of all results where the Euler formula has been applied to each separate component.

Another way of verifying the planarity of a graph is through checking its non planarity condition by applying Kuratowski's theorem.

Kuratowski's theorem: Any non planar graph contains a subgraph homeomorphic to K_5 or $K_{3,3}$.

K_5 is a complete graph with each node of degree 5; $K_{3,3}$ is a bipartite complete graph with each node of degree 3 (see Figure 4.10 and Figure 4.11 respectively).

Dual Graphs

The concept of a dual graph is also applied in spatial modelling. The most commonly known are the graphs that represent Delaunay triangulation and Thiessen polygons (Delaunay 1934, Thiessen 1911). Consider a graph G that has been embedded onto a plane. G has n nodes, m edges and f faces and G' has n' nodes, m'

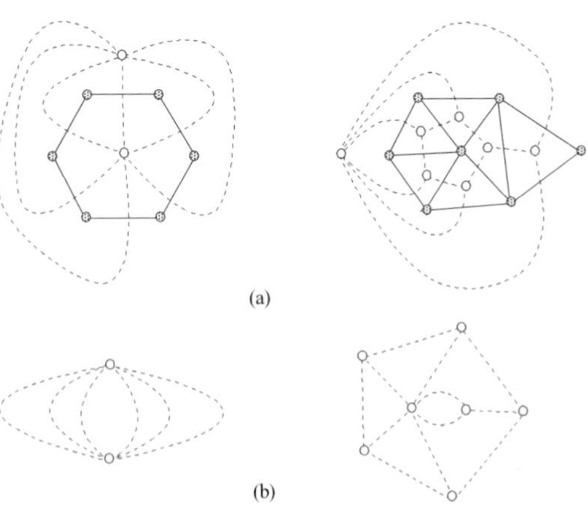

Fig. 4.13 (a) Graphs and their geometric duals that are isomorphic to graphs in (b)

edges and f' faces. G' is the (geometric) dual of G if:

(i) n' = f, m' = m and f' = n and such that each node n_i' of G' can be mapped inside each face f_i of G in a one-to-one correspondent manner, and

(ii) each edge e' of G' crosses a corresponding edge e of G, but no other edges of G, where e' joining two nodes of G', n_i' and n_j' that, respectively, lie inside two faces of G, f_i and f_j both adjoining e.

Observe also that any graph dual to G is isomorphic to G'. However, if G and H are isomorphic, G' and H' need not be isomorphic. Examples of graphs and their duals are illustrated in Figure 4.13.

4.6.8 Spatial Data Models in GIS

Based on the conceptual model of the real world presented in section 4.6.3, a number of spatial data models have been developed, for example DIME, ATKIS, TIGERS. In geoinformation science, data models may be categorized according to dimensionality, representation of space and thematic representation as explained in the following sections.

Multi dimension

If the aspects of reality to be modelled involve objects of various dimensions, the spatial model should provide the highest spatial dimension, so that all objects can be accommodated. For example, the data model may contain point, line, surface and body features whose dimensions range from zero to three. This kind of data model is regarded as multidimensional. The dimension number of the model is the highest dimension of the objects it can contain. The 2D data model is the most commonly used at present, but demands for 3D and 4D are increasing. The 3D data model may be regarded as equivalent to the static world, without any change over time. If the temporal component is also taken into account, the dimensionality of space is then increased to 4D, as found in the modelling of dynamic processes.

Tessellation

Tessellation is a complete and continuous subdivision of space into spatial units that may be of either regular or irregular shape. The model of the real world constructed by this approach may be regarded as a tessellation-based model. In 2D, we know several types of regular tessellations, for example squares, rectangles, hexagons, equilateral triangles and so forth. Squares are the most commonly used, for example for storage of digital images and surface models. While for irregular tessellation, triangles and Thiessen polygons are the most commonly used units. Triangular irregular networks are also frequently used for the storage of surface models, while Thiessen polygons are used to represent influence zones and the proximity of objects. Both irregular tessellations are important in this kind of work and are used to design data models. We therefore elaborate here on the irregular tessellations.

Two kinds of irregular tessellation are distinguished. The first is tessellation by complex geometry; the second relies on simplex geometry (see section 4.6.7 for the definition of simplex and complex).

Tessellation by Complexes

Tessellation by complexes is achieved by subdividing space into a set of cell complexes, for example, polygons and polyhedrons. The most important in this thesis is Dirichlet tessellation (Dirichlet, 1850), or Voronoi tessellation (Voronoi, 1908), based on the proximity of objects. If objects are represented by kernel points, a Voronoi region encompasses a set of points closer to a kernel point than to any other point in the set. This kind of tessellation can be applied to any dimension.

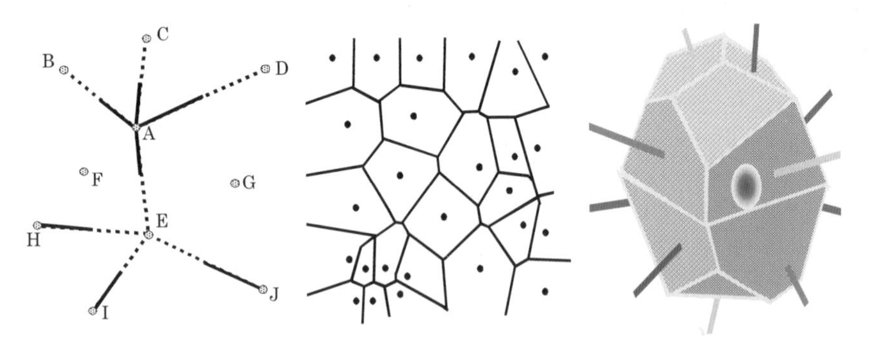

Fig. 4.14 1D Voronoi tessellation

Fig. 4.15 2D Voronoi tessellation

Fig. 4.16 An example of Voronoi polyhedron

Figure 4.14 is an example of 1D Voronoi Tessellation applied along any straight line connecting two points. The influent zone of each kernel point covers the distance from the point up to the middle of each line emanating from the point, and connecting to the other point. For example, between the points A and C connected by a straight line, the influence zone of A is expressed by the dark, solid part of the straight line AC, while the influence zone of C is expressed by the dotted part. Figure 4.15 is an example of a 2D Voronoi tessellation, also known as Thiessen polygons (Thiessen, 1911). Voronoi polygons are shown in thick lines, while kernel points are shown as small black circles. A 3D Voronoi polyhedron may look like Figure 4.16.

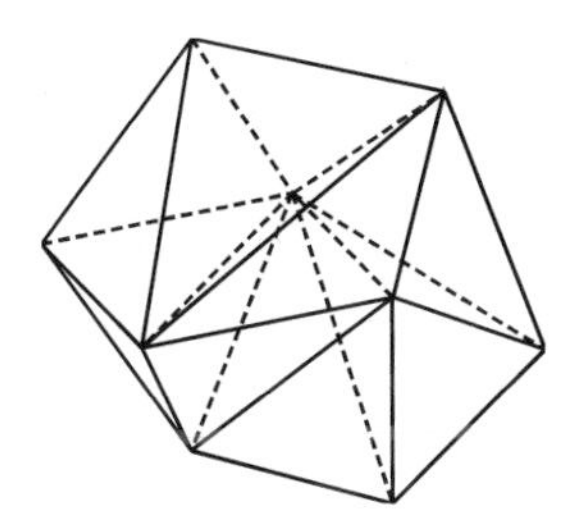

Fig. 4.17 A network of tetrahedrons.

Tessellation by Simplices

Tessellation by simplices is done by subdividing space into a set of simple objects, for example triangles and tetrahedrons. For 2D, the tessellation by triangles is known as a triangular irregular network. Its extension to 3D is a tetrahedral network (Figure 4.18). Different kinds of TINs are distinguished by the extent to which they have been considered important in this thesis (see chapter 6 for more details). Delaunay triangulation is the most popular method for TIN construction. A large number of publications have already discussed this topic in depth (Delaunay, 1934; Sibson, 1978; Tsai, 1991). With respect to graph theory, Delaunay triangulation and Thiessen polygons (2D Voronoi) are geometric duals and each can readily be derived from the other. Figure 4.17 shows both Thiessen polygons and a Delaunay triangulation in solid lines and dotted lines respectively. Likewise, Delaunay tetrahedral network and Voronoi polyhedrons are geometric duals. Figure 4.16 shows lines emanating from the node inside the Voronoi polyhedron; they are the edges

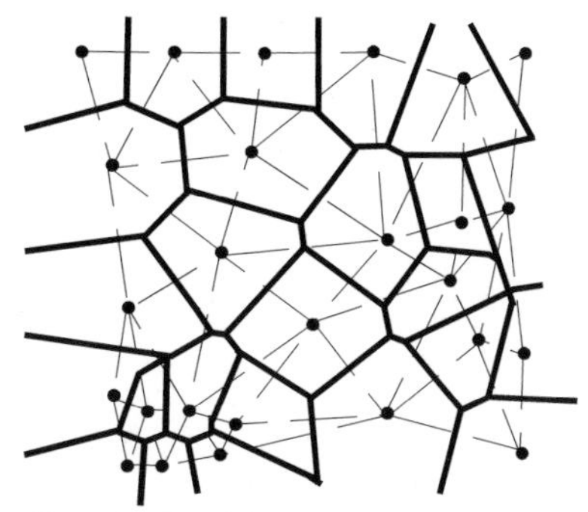

Fig. 4.18 Thiessen polygons and Delaunay triangulation.

of tetrahedrons.

Delaunay triangulation has been used to facilitate spatial computation, for example in DTM for the derivation of contour lines, interpolation of height at a given planimetry, and computation of slope and aspect.

Single-theme and Multi-theme

Two kinds of data models can be distinguished based on the representation capability of a geometric element of the data model. If a geometric element is part of the representation of only one real world object (of one thematic class), the data model is regarded as a single-theme data model (Molenaar, 1989). If a geometric element is part of the representation of more than one real world object (of more than one thematic class), the data model is regarded as a multi-theme data model (Kufoniyi, 1995).

Single-theme

Molenaar (1989) suggested a spatial data model for vector representations of geo-spatial objects. What he called the formal data structure (FDS) of a single-valued vector map (SVVM) geometrically abstracts spatial objects are objects such as monuments, roads, rivers, forests and parcels to points, lines and areas. A terrain feature is described by an identifier, its geometric primitives (arc and nodes), and its (single) thematic class. Figure 4.19 shows the graph representation of this conceptual model. Although the coordinates of every node can be 3D, the spatial data model provides only 2D topology. The FDS approach provides a highly disciplined approach to geoinformation modelling. The data model is presented by a diagram and a set of conventions that provide the rules for modelling. The conventions help to prevent confusion during modelling and subsequent processes when operations need to be applied to the data set and facilitate formulating consistency rules for updating a geo-spatial database.

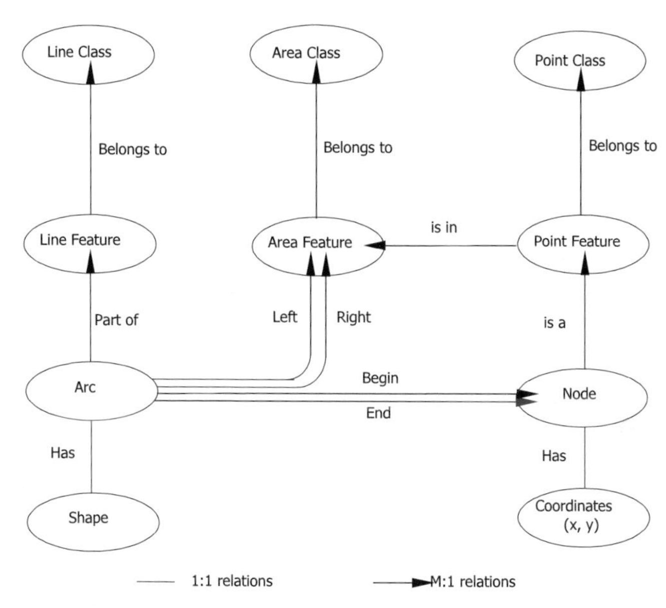

Fig. 4.19 A single-theme data model (after Molenaar, 1989).

Multi-theme

A multi-theme (or multi-valued) vector map (MVVM) refers to the integrated representation of geo-spatial objects from more than one theme in the form of point, line and area feature types in a database. A graphical representation of the multi-theme data model is shown in Figure 4.20.

The multi-theme data model extends the concept of single-theme by adopting the idea that many spatial objects can overlap by sharing the same subspace. In the data model they can share the same geometry, because reality can be viewed differently by different observers, or for different purposes. For example, soil and water can coexist at the same location during flooding, or for the analysis of moisture content of the soil. The multi-theme data model provides additional geometric data types to group the geometric primitives shared by more than one feature. Since this group belongs to different features with different themes, it is said to have a heterogeneous (thematic) property. The feature itself belongs to only one theme, so it is said to have a homogeneous (thematic) property.

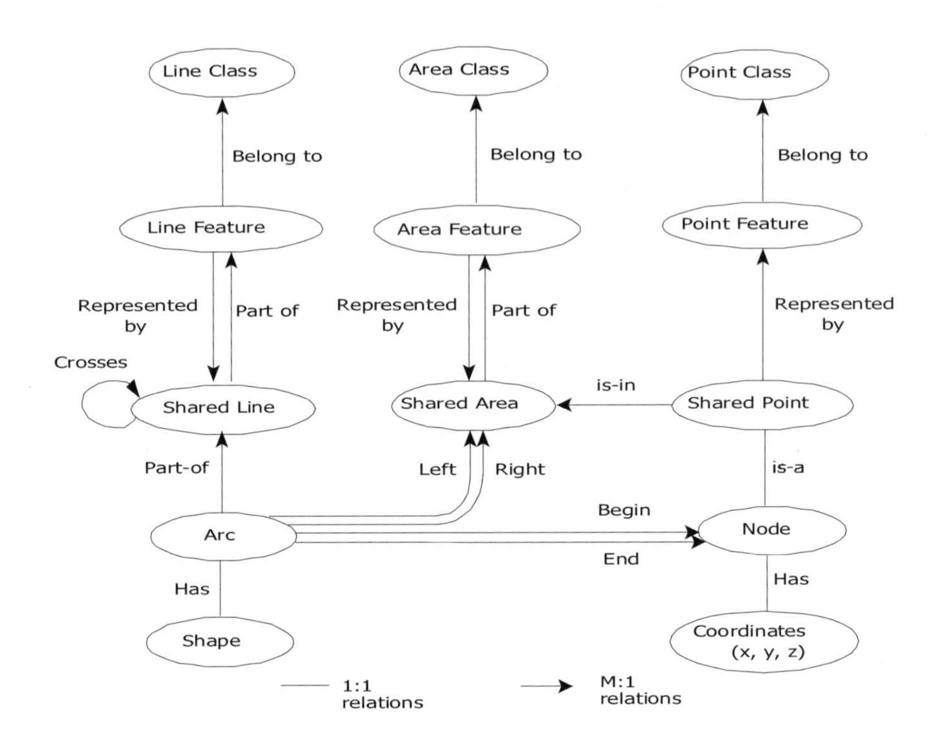

Fig. 4.20 Data Model for Multi-valued Vector Maps (Kufoniyi, 1992)

4.7 Logical Design of Geo-spatial Model

The logical design defines all the data elements needed for the representation of each spatial object. It outlines the method for the translation of the data model from the conceptual level to a logical level. Three kinds of logical data models are commonly used in information science: hierarchical, network, and relational. The hierarchical structures are those tree-like structures where a record is linked to another record via a 'parent and child' relationship. Based on the strict ordering of relationships, the structure is an acyclic directed graph in which no recursive relationship is allowed.

Less strict than the hierarchical is the network structure, because it allows any kind of link between data elements, provided there are elements at both ends of the link. The operations using network data structure are fast and efficient, but the design, implementation and maintenance of this data structure are rather difficult.

The relational structure is more closely related to natural ways of thinking. Regardless of performance, relational structure offers the quickest and the easiest way to logical design. Another important logical design which has become popular during the last decade is based on the object-oriented approach. The object-oriented approach extends the hierarchical and network structures by encapsulating related operations as additional attributes within each record. Logical design based on the object-oriented approach is rather involved, since we must also take into account the related operations. Only the relational and object-oriented approaches are reviewed here, since they are used in the design of a unified data structure (UNS) in later chapters.

4.7.1 Relational Approach

A relational structure is a collection of relationships between data elements representing aspects of reality based on set or relational algebra (see Date, 1986; Howe, 1989; Martin, 1983). The logical design of a spatial model based on the relational approach can be achieved by organizing the representations of spatial objects and relationships between them into a set of tables consisting of rows and columns (that is, records and fields). Figure 4.21 is an example of the relational data structure of the SVVM shown in Figure 4.19.

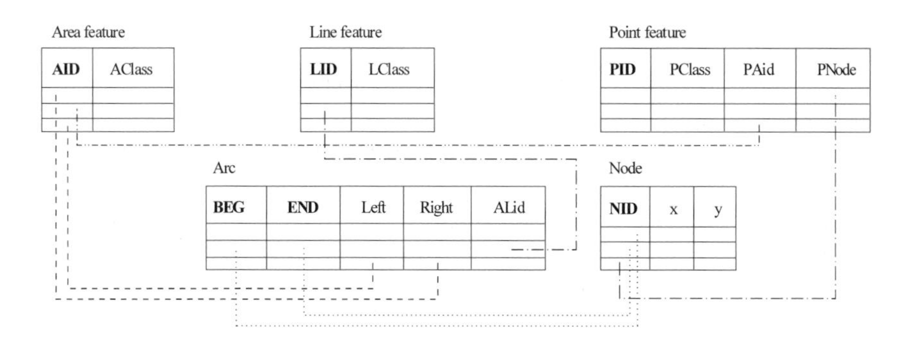

Fig. 4.21 Relational data structure of SVVM (adapted from Kufoniyi, 1989)

Point, Line and Area feature tables contain relationships between feature level and class level. Point, line and area features are represented by identifiers, that is, PID, LID, AID, whereas classes are represented by PClass,

LClass and AClass in their respective tables. Neighbourhood relationships between each arc (indicated by BEG and END nodes) and its left and right area features are presented in the Arc table. The field ALid in this table indicates to which line feature each arc belongs.

Since the relational data structure consists of several tables, a Cartesian product— a set of possible combination of rows of different tables— is used to further derive relationships between data elements across the different tables being joined together. For example, to know the coordinates of the beginning and end nodes of an arc, the tables Arc and Node are joined together. The search is carried out on the Cartesian product of these two tables to find the match between the values of BEG or END with the values NID. If the match is found, the coordinate X and Y are obtained.

It is important for this kind of data structure for all relationships to be optimized to prevent anomalies occurring with updating. Normalization is the optimization process that can be carried out stepwise, that is, from the first to the fifth normal form. In general, the third normal form is found to be acceptable. Several approaches for normalization exist, for example, non loss decomposition, entity-relationship approach (Chen, 1983), dependency diagram (Smith, 1985). The last mentioned is considered to be the most intuitive approach.

Normal Forms

The first normal form (1NF) is obtained after eliminating repeated groups. The second normal form (2NF) is obtained by elimination of the non identifier attributes which are not functionally dependent on the whole key. The third normal form (3NF) requires the elimination of all the functional dependency between non key attributes. The fourth normal form (4NF) deals with multi-valued dependency that occurs when an attribute value can be inferred from another record. The fifth normal form (5NF) deals with the joint dependency that occurs when some facts are stored twice in the same table. This situation may be regarded as a result of joining two pairs of attributes in a record with one attribute in common. Hawryszkiewyc (1991) discusses and provides examples of this issue.

Smith's Normalization

Normalization is commonly achieved by decomposing a preliminary table into first, second, third, fourth, and fifth normal forms; in this way, several normalized relationships in the form of tables are obtained. This seems, however, to be a very tedious task. Smith (1985) proposed a more attractive approach, whereby tables are composed from a dependency diagram. If the procedure is followed correctly, the database tables obtained are then fully normalized. The steps can be summarized in five phases:

1. Identifying all the data elements
2. Constructing dependency statements
3. Mapping from dependency statements to dependency diagram
4. Composing relational tables from the dependency diagram
5. Improving the handling of the relational table by introducing a surrogate key if necessary.

Roessel (1986) has demonstrated the applicability of this method to a spatial database. The same approach has also been used by Kufoniyi (1989), Bouloucos et al., (1993), Ayugi (1992), Pilouk and Tempfli (1992), and Chhatkuli (1993) because it helps clarify and creates more understanding about spatial relationships. The normalization would also be useful for the transition into object-oriented database structure, where the same kinds of relationships have to be represented. This approach will be described in later chapter.

4.7.2 Object-oriented Approach

The object-oriented approach was developed about two decades ago, starting from the developments of the object-oriented programming languages Simula 67 (Dahl et al 1970) and Smalltalk (Goldberg and Robson 1983). Object-orientation has been built up from the idea of encapsulating data and operation, and processing the data together. The confusion about which data must be processed by which procedure is thereby avoided. Furthermore, computer codes need to be reused, since an algorithm may be applied to different types of data in different applications. The object-oriented approach provides mechanisms allowing economic reuse of a computer code by extending the code to accommodate additional types of data without a major reprogramming effort. Moreover, it also provides abstraction mechanisms to natural model spatial objects.

Encapsulation

This is a mechanism tying together the attributes describing the state of the object and the operations retrieving information about the object, or changing the state of the object. These operations are also known as an object's behaviour, method, and dynamic properties. Encapsulation provides the object with the control determining which attributes and behaviours would be private properties and which is accessible to the public. Encapsulation can ensure that an operation is applied to the right object, thus preventing ambiguity during operation. This is also known as a type-safe operation.

Classification

Each object needs to be organized into a certain class. Objects with the same kind of properties and behaviours should be placed in the same class. Each object is said to be an instance of a class. A class is a place for defining the specification of an object. A class is said to be an abstract data type (ADT) in the sense that we have an opportunity to create new data types that fit our abstraction about a real world object.

Inheritance

When an object has been defined, the inheritance mechanism permits the propagation of the properties and behaviours to lower level objects in the same hierarchy. The propagation from one object to another object is allowed if the two objects have 'parent' and 'child' relationships. The child inherits all the properties and behaviours from the parent through a class mechanism. The inheritance is activated by deriving a new class from the existing class. If the new class is derived from only one existing class, it is known as single inheritance. It is called multiple inheritance if the new class is derived from more than one existing class. The inheritance makes the existing class reusable and thus significantly saves time in redesigning a new class, provided that the new and existing classes have something in common. If the inheritance is defined at the logical design stage, it is called static inheritance. If the inheritance needs to be defined during the construction of spatial model, it is called dynamic inheritance (Weiskamp and Flamig 1992). Dynamic inheritance is needed when multiple representation of an object is required and the representation is not known prior to the construction of the spatial model. For example, a city may be represented as a point, or as an area feature highly dependent on the user. To make the logical design flexible, the type of representation can only be decided upon during the construction of the spatial model.

Generalization and Specialization

If two or more classes have many properties and behaviours in common, a more general class can be created to become their parent. Conversely, when deriving many new classes from an existing class, each class may have, in addition to the parent, more specific properties and behaviours. The former scheme is generalization and the latter is specialization. These two mechanisms permit the streamlining of classes, making them easier to maintain. Generalization and specialization create a class hierarchy using an inheritance mechanism. The generalized class exists at the higher levels of the hierarchy, while the more specialized class exists at the lower levels of the same hierarchy. The complexity of objects increases from higher to lower levels of the hierarchy.

Aggregation

The design of a class needs to include many data types and behaviours. An aggregation process is involved. Each data type aggregated into a class may be either system or user defined. Different existing classes may be aggregated to build up an aggregated class. Objects of existing classes are components of a composite object in the aggregated class. The relationship between each component and the composite object is of the type part-of. The aggregated class does not inherit properties or behaviours from its component classes; neither is it a parent of each component class. The properties and behaviours of each component are normally inherited from their respective parents. Relationships between components are well defined. Each component is necessary for the constitution of the composite object. The composite object cannot be independently constructed. The aggregated class should have its own behaviours. The behaviours of the components are normally suppressed by the aggregated class.

An example of an aggregated class is the object 'car'. It is composed of objects from the classes wheels, engine, steering wheel, and so forth. A car does not inherit properties or behaviours from, say, the wheels or the engine. A car resembles neither a wheel, nor an engine. A car is not complete if the component wheel, or engine, is missing.

An example in GIS can also be given. A line feature is a composite object consisting of a list of nodes. These nodes must be arranged in proper order; each pair of nodes defines a straight line segment that constitutes the geometry of the line feature. If the nodes are badly arranged, or if one of them is missing, the line feature cannot be correctly constructed.

Association

In many cases, an object can be a container, or a collection, of many other objects. The association defines a membership relationship between a group, or a set, of lower level objects (members) with a higher level object—the container. Relationships among member objects are not always clearly defined, but the relationship between each member object and the container must be clearly defined. The association is useful for the design of container classes, for example, array, linked-list, stack, or queue. The member objects need not be of the same class. Although the container in computer science is commonly used to contain the same type of objects, as for example in an array of integers, in reality a container can contain many different objects. Two kinds of container exist: homogeneous and heterogeneous. A container is complete in itself and can be independently constructed without the member objects. In other words, a member object need not constitute, nor be part of, the container. For example, an arc container may be an array containing a set of arcs. This set of arcs does not constitute the container. Whether or not this set, or subset, of arcs constitutes a line feature is not important for the container. The container's only interest is whether an arc resides within the container or not. With respect to this point, a container's own behaviour lies very much in the management of the membership of objects. The behaviours of each individual member object are normally suppressed, while the group behaviours are promoted.

The use of association and aggregation are easily confused, because both of them relate to many objects at a lower level. We can consider the case of a line feature as an example to decide whether association or aggregation of a set of arcs is more suitable. In the spaghetti model, where the relationship between any two consecutive arcs of a line feature is not considered important, the line feature may be treated as a collection of arcs. Here, the association may be adequate. In a topological model like SVVM, where traversing along the line feature to support network analysis should be possible, aggregation is more suitable because the relationships between two consecutive arcs are also considered important.

Polymorphism

Polymorphism allows two or more classes to use the same attribute name without confusing about the class to which the attribute belongs. This is done by associating an object with its attribute. For example, an area feature and a line feature can have the same attribute 'length.' Length, as an attribute of the area feature, is the perimeter of the area boundary, whereas it is simply the length of the line feature. Another kind of polymorphism is called parametric polymorphism, whereby the class name is used as parameter to make the general design specific (Pohl 1993). The general design of a class is known as the template class. General dynamic behaviour can also be provided as a template function.

4.8 Summary

In this chapter, we have reviewed some fundamental and theoretical concepts related to spatial modelling. These concepts are important for the understanding and development of a spatial model in different phases. Two phases of spatial modelling are distinguished: design and construction. The maintenance phase, keeping the constructed spatial model valid, however, is not covered in detail here. On the basis of the scope of involvement by different disciplines in the design of spatial model which Molenaar (1994b) describes, only the conceptual and logical design are within the scope of geo-information theory. The detailed internal design is left for computer scientists; it is therefore outside the scope of this book.

For the conceptual design, mathematical concepts about space and relations, that is metric, order, and topology are considered important. Concepts of simplicial complexes and the theory of graphs are used to represent real world objects in the spatial model. Existing spatial data models resulting from different conceptual design approaches, that is, SVVM, MVVM, irregular tessellation by simplices, are considered fundamental to the modelling approach used here. For the logical design, relational and object-oriented approaches are fundamental to the design of spatial data structures.

The elements of spatial theory provide a sound basis for the design of an integrated geo-spatial model that permits the representation of both determinate and indeterminate spatial objects. SVVM, for example, is a model representing determinate objects. Simplicial complex and Delaunay triangulation provide fundamental concepts for the representation of indeterminate spatial objects. The mathematical concepts provided in this chapter can be used to support the conceptual design for both FDS and integration approaches. FDS can be formulated using cell complexes, or simplicial complexes. For example, a face is a 2-cell complex which can be decomposed into a 2-simplicial complex by using the Delaunay triangulation method. The same mathematical concepts can be used to evaluate existing data models, whether or not the provision of the elements for the integrated modelling of reality is sufficient.

Chapter 5 THE CONCEPTUAL DESIGN

This chapter elaborates on the development of an integrated vector data model as a basis for the implementation of a core database in a 3D geoinformation system. The tessellation approach is used with the aim of extending the query space as offered in the object-based approach. The tessellation approach accommodates a wider range of complex spatial analyses that involve both computation (for example, interpolation, slope, aspect, visibility, shading, surface area, volume) and simultaneous topological navigating in the database (for example, selection, indexing, sorting). The combination of irregular tessellation and the application of topology, as found in FDS and described in chapter 4, can offer such an opportunity. They are therefore adopted to strengthen integrated 3D modelling. We first elaborate the 2.5D model which integrates terrain relief and terrain features. Since this data model is mathematically sound, it can be readily extended to cover aspects required for 3D modelling. To ensure forward compatibility, so that the 3D spatial model can be incorporated into a higher dimensional spatial model in the future, a generalization for nD is also presented. The properties of the integrated data model are discussed within the framework of simplicial complexes and graph theory. This discussion leads to the definition of a simplicial network as well as its Euler characteristics.

5.1 TIN-based (2.5D) Data Model

The first stage of development of the integrated data model aims at a structural integration of the representation of terrain relief and terrain features, which is regarded as 2.5D. The terrain relief represents the geometry of the earth's surface, where terrain features are 2D representations of spatial objects. Terrain relief and terrain features information in the forms of DTM and GIS, respectively, are familiar tools for solving spatial related problems, decision making and mitigating hazards, such as erosion and landslides, agricultural land reformation. Since the integrated database should fulfil the requirements of a typical DTM and GIS (see chapter 4), it must ensure all the functionalities of the two systems.

To design the integrated terrain relief and features data model, TIN and 2D FDS have been selected, for the reasons indicated below.

1) Efficient interpolation

Two main underlying principles adopted in the design of this data model are proximal ordering and decomposition into primitives. Digital terrain relief modelling requires interpolation, which in turn requires the proximal relationships among the given points to be known. As DTMs are in most cases based on non-gridded data, adjacency can best be expressed by triangulation. Triangulation of surveyed points in order to interpolate contour lines for topographic maps has in fact been applied long before the advent of computers. Thiessen (1911) polygonal tessellation and its geometric dual, the Delaunay triangulation (see chapter 4, Pilouk and Tempfli 1992), are of special interest. This is through their establishment of a natural neighbour structure facilitating the interpolation process where different methods of interpolation (see Tempfli, 1982) can be applied.

2) Fidelity and embedding of constraints

One way to increase the fidelity of surface representation is to incorporate skeleton data, such as ridge and drainage lines, in the DTM. Different approaches have been suggested, for example, by composite sampling — the combination of selective and progressive sampling (Makarovic, 1977) and constrained triangulation (Pilouk, 1992). The latter is capable of maintaining skeleton data without losing their original geometry. Constrained triangulation permits the embedding of such geometric components of skeleton features as the components of triangles; for example, a line feature can be decomposed into a series of straight line segments and be embedded as edges of triangles. The same approach as for the skeleton data can be applied to embedding geometric components of other terrain features in the TIN structure without losing their original shape as obtained from observation or measurement. These features represent the human knowledge about the aspects of reality which should be recorded correctly into the database and so their information must be maintained. As a by-product, the fidelity of surface representation increases when more terrain features are incorporated into the TIN. This capability of embedding the terrain features is one of the most important aspects in the design of the integrated data model.

3) Locality

Locality is an important aspect of a very large data set where a large amount of time is obviously required to process the whole data set, as in data retrieval, calculation, and updating. As pointed out in the literature, TIN permits local editing and updating of elevation data without the elaborate re-interpolation necessary in a grid-based structure (see Fritsch and Pfannenstein, 1992a). The local editing of TIN structure that only involves necessary data has been demonstrated in (Jackson et al., 1989, Pilouk et al., 1994; Midtbø, 1993).

4) Convex shape

A triangle is a convex polygon. Its simplicity reduces uncertainties and consequently requires less testing, thereby offering significant advantages in many graphic and geometric operations where fast computing speed is crucial. Some examples of these operations are the inclusion of a point in a polygon, as in colour filling, and rasterization, intersection with a line and a polygon, as in hidden-line/surface removal, and finding the direction of the surface normal vector to determine the reflection of light and surface visibility.

5) Finiteness and adaptability

The number of vector elements that are the components of a triangle is fixed. It is, therefore, easy to control the consistency of its network. Nevertheless, a triangle's finiteness does not prevent the adaptability of the network to the terrain roughness, because the network of triangles can be densified without limit. The irregular shape of a triangle permits its adaptation to the irregular distribution of observations or measurements.

6) Compatibility with FDS

TIN has typically been used for the representation of surface geometry. Since it is vector-based, creating the links between geometric components and features that have links to thematic components in a similar way to that defined in FDS is highly feasible (Molenaar, 1989).

Based on the above reasons, the design of the integrated terrain relief and features data model is as shown in Figure 5.1.

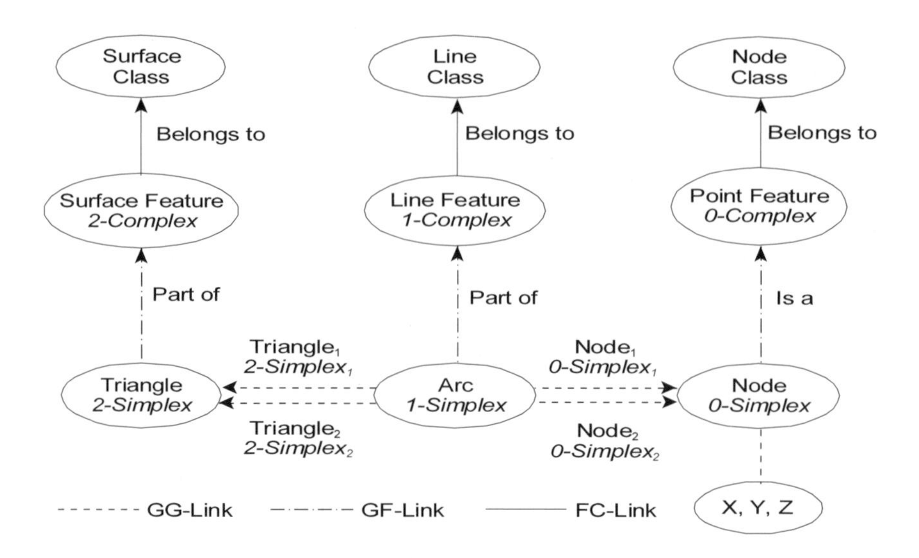

Fig. 5.1 The proposed integrated 'DTM-GIS' data model.

5.2 Properties of the TIN-based Data Model

The TIN-based integrated data model can be considered as the extension of the 2D FDS of SVVM on the geometric level. The geometric primitive triangle is added, and a data model obtained which serves the purpose of the unified handling of all surface related data. In comparison with 2D FDS, some links presented in 2D FDS are redirected in order to streamline the links. An area feature is no longer linked to arc directly, but to its geometric primitive triangle. As a result of the decomposition, an area (surface) feature then consists of one or more triangles. Arcs are linked to triangles through the left and right links, labelled in Figure 5.1 as Triangle1 (2-$Simplex_1$) and $Triangle_2$ (2-$simplex_2$), respectively. The vertices of a triangle can be found through the arc-triangle and arc-node links. Each node is represented by one coordinate triple which consists of one X, one Y, and one Z. Since all nodes are assigned 3D coordinates, a plane equation can be derived from the three vertices of a triangle. This allows further derivation of information relating to terrain relief, such as elevation at any point, slope, and aspect. Additionally, the real-time visualization is no longer limited to orthogonal views (the traditional map). Popular visualizations, such as perspectives and stereo views - even with shaded relief- surface illumination or texture mapping, can be generated more efficiently, and can be combined with a cursor to access information from the '3D graphics.'

Following the FDS approach described by Molenaar (1989), the integrated terrain relief and features data model consists of three levels, described in a top-down manner:

1) Class level
This level consists of thematic class data types that maintain information related to the application, or the manner in which the features described in the next level will be used. For example, a line-feature as a road, a surface-feature as an industrial zone, a point feature as a city. The classes are mutually independent.

2) Feature level
This level consists of three feature data types: point-feature (0-complex), line-feature (1-complex), and surface-feature (2-complex). Each level implies the type of geometry to be used for its geometric representation. This level provides the interface to the user. Each feature type maintains a feature-class (FC) link to exactly one class data type.

3) Geometry level

This level consists of three geometric data types, that is, node (0-simplex), arc (1-simplex) and triangle (2-simplex). Each of them maintains a geometry-feature (GF) link to a feature it composes, so that a node may represent a point-feature, an arc (which is a straight line) may be a part of a line-feature, and a triangle must be a part of a surface-feature (Figure 5.1). Within the same level, there are also geometry-geometry (GG) links between two related geometric primitives; that is to say, an arc has two nodes - a beginning and an end - and it has two triangles - left and right. The node type maintains the georeference to the external space in the form of a coordinate tuple.

Note that the links represented as arrow-headed lines in Figure 5.1 only indicate that those links are possible, but not necessary. For example, some arcs are not part of any line-features and some nodes are not point-features. Nevertheless, every node must be a vertex of a triangle and every arc must be an edge of a triangle. In comparison with FDS, a node may be isolated and an arc need not be an edge of a polygon. This interpretation will become clear in chapter 6, where the mapping into a relational database structure is explained in detail.

In terms of simplicial complexes (Egenhofer et al 1989), this data model consists of 0-simplices (nodes), 1-simplices (arcs), and 2-simplices (triangles), with the smallest data elements of 0, 1, and 2 dimensions respectively.

With respect to the transition from an object-based data to a tessellation-based model, the following requirements defining a decomposition scheme must be fulfilled:

1. A surface-feature (2-complex) is composed into a set of triangles (2-simplicial complexes).
2. A line-feature (1-complex) is composed into a set of contiguous arcs (1-simplicial complexes) that are in fact triangle edges.
3. A point-feature (0-complex) needs no decomposition; it is simply a node (0-simplex) that is a triangle vertex.

Following the syntactic approach of Molenaar (1994b), the above decomposition scheme can be mathematically described as follows:

$$\text{PartN}\ [S_N , C_N]$$

where,

PartN = part of relation at dimension N
S_N = simplex of dimension N
C_N = (complex) object of dimension N

A decomposition process that guarantees the above requirements must be made available to facilitate this. Delaunay and constrained triangulation can serve this purpose. The process is described in more detail in chapter 7.

By assigning 3D coordinates (x, y, z) to every 0-simplex, the mapping of this model in 3D metric space becomes meaningful in, for example, visualization, or the calculation of slope or surface area. Consequently, various topological operations and the derivation of topological relationships can be readily performed by using the basic binary relationships (Egenhofer et al 1989), because all objects are said to be decomposed into minimal spatial objects of their dimensions.

The data model presented in Figure 5.1 only accommodates single-theme GIS, which does not permit the sharing of the spatial region of different objects of the same dimension in a database. So the next step of the development of the integrated data model is to add the capability of handling multiple themes stored in the same database. The multi-theme concept developed by Kufoniyi (1995) described in chapter 4 can be adopted. By adding a sub-feature level that consists of three new data types between geometry and feature level, the more integrative data model as presented in Figure 5.2 is obtained (Pilouk and Kufoniyi 1994).

This further developed data model enables the sharing of spatial regions and the simultaneous representation of terrain relief. It is achieved by redirecting the links from geometry level to the sub-feature level instead of direct links to the feature level. So, instead of being an explicit part of the feature, each geometric primitive is then a part of a sub-feature and thus indirectly a part of a feature. The sub-feature is in fact an aggregation of the geometric primitives. The sub-feature level ensures the connection to the theme by maintaining the 'part of' links to the data types at the feature level. Observe that the sub-feature data types represent overlapping spatial regions being shared by more than one feature of different themes. Each sub-feature can then be a part of more than one feature and therefore has a *heterogeneous* thematic property, while the feature itself is *homogeneous,* because it represents only one object of reality that has a unique property across its spatial extent (see chapter 4). A feature still belongs to only one thematic class, so it does not lose its property of being single-valued. This convention also suits data acquisition which is usually carried out per theme.

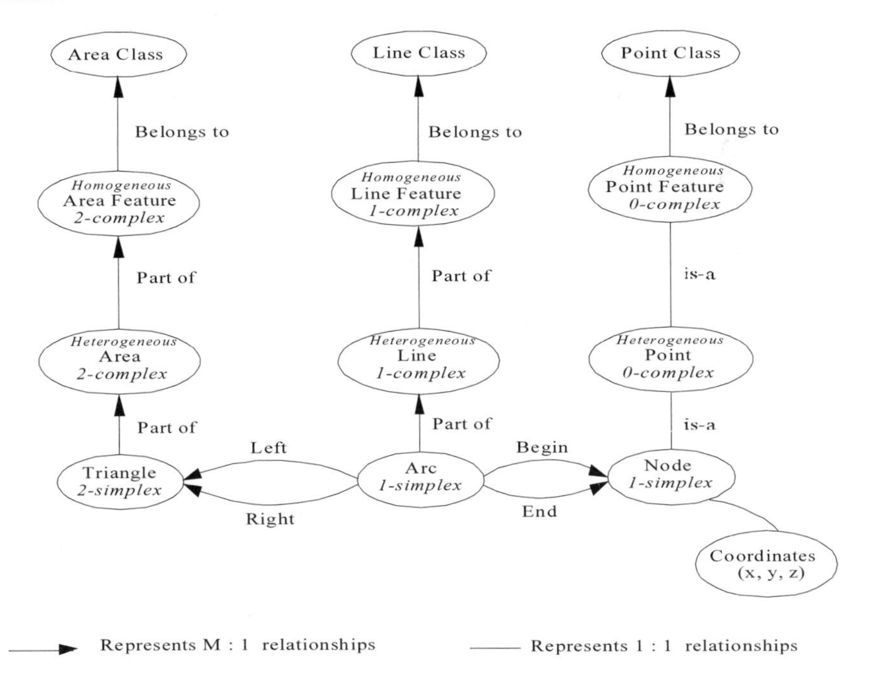

Fig. 5.2 Integrated data model for DTM and multi-theme GIS (after Pilouk and Kufoniyi, 1994)

The model implies the following decomposition scheme:

1. Each area feature is decomposed into a set of subareas; each subarea is still a 2-complex and is therefore further decomposed into a set of triangles.
2. Each line feature is decomposed into a set of lines; each line is still a 1-complex and is therefore further decomposed into a set of arcs.
3. No decomposition is needed for any point feature; it is considered as a 0-complex and a 0-simplex (that is to say, a node) at the same time.

The concept of Delaunay and constrained triangulation can still be used for the decomposition into primitives, but, only after the decomposition of features into sub-features for which the typical overlaying process in GIS can be used.

5.3 TEN-based Data Model

The TIN-based data model presented in section 5.1 is limited to applications that consider single-valued surfaces (see Figure 1.1). It has no capability to serve applications that need to deal with multi-valued surfaces, or solid bodies. Applications in geology, geo-science, architecture, civil engineering, urban planning, facility management and environmental monitoring all require full 3D spatial information, in which an integrated data model that can represent multi-valued surfaces and solid objects is needed.

The TIN-based data model has to be extended to facilitate handling of 3D objects in particular, in order to stretch the capability of the integrated data model in both dimensionality and computability. The triangular network can be generalized into a tetrahedral network. Delaunay triangulation can also be generalized for tetrahedronization (see chapter 7).

As discussed earlier, the general properties of a tetrahedron are the same as a triangle's; each is a simplex of its dimension and convex. Some important properties of their networks, for example, locality, fidelity and capability of embedding features are also similar. The latter indicates that the geometry of the features can also be maintained within the tetrahedral network (TEN), which means that TEN also has the capability of maintaining human knowledge about the real world characterised as follows:

1. A body-feature is a contiguous set of tetrahedrons that is a subset of the TEN.
2. A surface-feature is a contiguous set of triangles that are faces of tetrahedrons.

3. A line-feature is a contiguous set of arcs that are edges of tetrahedrons and triangles.
4. A point-feature is a vertex of at least one tetrahedron.

The above statements may be treated in addition to the set of conventions for 3D FDS; for example, self-overlapping or self-intersecting of a feature is not allowed.

For the aspect of interpolation, the bivariate interpolation methods, for example, the weighted average (Tempfli, 1982), can be generalized into trivariate; that is to say, values are estimated such that $p = f(x, y, z)$.

For better understanding, we shall first discuss the single-valued variant as shown in Figure 5.3. Compared with the TIN-based data model in Figure 5.1, the main differences (apart from the number of data types) are the GG-links between the arc and the triangle. In Figure 5.1, the arc maintains the left and right links to the triangle, while in Figure 5.3, the triangle provides three links to the arc. These three links differentiate arcs as three triangle edges. This differentiation is needed to normalize the many-to-many link from triangle to arc into three many-to-one links; that is to say, a triangle has arc X as 'edge-1', while arc X can be an 'edge-Y' (Y = 1 to 3) of many triangles in 3D space. The left and right links from the arc to the triangle are eliminated since in 3D, more than two triangles can share one arc. The geometric type 'triangle' is comparable to 'face' in 3D FDS. However, each triangle has only three edges and three nodes, so it is not difficult to determine the triangle's orientation by ordering the three edges. This order can be subsequently recorded for each instance of the triangle data type and does not consume additional storage. Such order makes it possible to omit the edge data type that keeps the information about face orientation that is important in 3D FDS, where the number of edges and nodes of the face data type can be varied. Storing the direction of each arc in the edge data type helps avoid the determination of face orientation, which may require considerable processing time. The orientation of each triangle helps to further determine the first and the second tetrahedrons situated on the positive and negative normal of the triangle respectively. (These are comparable to the left body and right body of a face in 3D FDS).

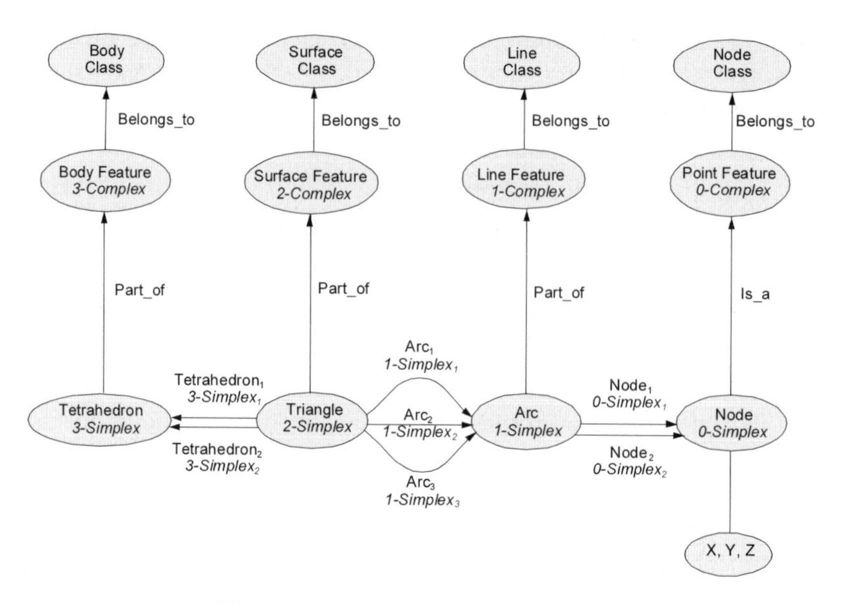

Fig. 5.3 Tetrahedron-based data model

A more precise description of the TEN-based data model can be given in terms of FDS together with simplicial complexes.

1. An instance of the node (0-simplex) data type has x, y, and z coordinate types as its attributes. It may be a part of an instance of a point feature (0-complex) type.
2. An arc (1-simplex) data type is defined as a straight line; it is therefore composed of only two instances of the type node, one on each end. It may be defined as a part of an instance of a line (1-complex) feature type.
3. A triangle (2-simplex) data type is composed of three arcs. It is shared by two tetrahedrons (3-simplices), one on each side of its plane (called the 1st and 2nd tetrahedron respectively). A triangle may be a part of a surface feature.
4. A tetrahedron (3-simplex) is a part of a body (3-complex) feature.

Observe that the tetrahedron data type does not carry any geometric description (triangular faces, edges, vertices), since its components can always be found from the geometric links with the triangle data type, being either the first or the second tetrahedron of a triangle (comparable to the left or right body in FDS terminology).

To extend from single-theme to multi-theme, we must augment the sub-feature level by one more data type - the sub-body - and redirect all necessary links in the same way as for the TIN-based data model. The TEN-based version that is capable of handling 3D objects with multiple thematic representations is shown in Figure 5.4.

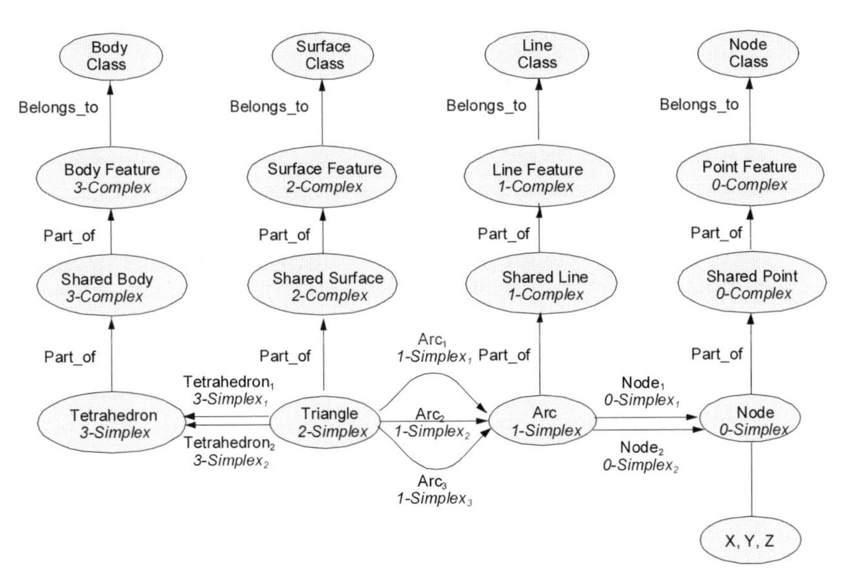

Fig. 5.4 Multi-theme tetrahedron-based data model.

5.4 Generalized n-dimensional Integrated Data Model

Observing the similarities between the data models shown in Figures 5.1, 5.2, 5.3 and 5.4, we can establish a general concept of an integrated data model based on irregular tessellation which may be useful for the study of multi-dimensional spatial information. In the different stages of the development, proceeding from TIN-based to TEN-based, both single-theme and multi-theme can be formalized. Theoretical support to this generalization is given by:

1. The FDS, which clearly represents relationships between the real world objects and how components of their representations are related in the spatial model.
2. The simplicial complexes, which help simplify the spatial objects and systematically and consistently map them into the representations in the model.
3. Graph theory, which can be used to rigorously describe the representations and which also provides the mechanism to ensure the integrity

of the overall representations, that is, the irregular network in this case.

An important benefit of having a theoretical basis for the tessellation-based integrated data model as a basic standard is that the compatibility across different dimensions can be established, thus:

1. It is more convenient for the user to decide what kind of data model to select; single-theme or multi-theme, and in what dimension. The user can instantiate a requirement as an input parameter to the generic data model and obtain the suitable model for the application. Users need not worry whether the databases at hand are based on or limited to a certain dimension. The generic data model makes possible the handling of data across different dimensions.
2. The user can navigate in different databases from one dimension to another dimension via the compatible links in various network structures, for example, from body, tetrahedron, triangle, arc, node and coordinates, provided that other databases also adopt the generic data model. In this sense, the generic data model can be regarded as dynamic.
3. The more efficient organizing, sharing and exchange of data and the elimination of disparity and redundancy lead to significant cost reductions. Avoidance of duplicate data collection is also feasible if the core database is widely accessible (Shepherd, 1991).

Prior to the design of the generic version of the integrated data model, a set of definitions must first be introduced.

5.4.1 The Definitions

We limit our consideration to geometric modelling and recall the mathematical description of spatial objects following the theory of combinatorial topology described in chapter 4. This theory classifies spatial objects according to their spatial dimensions defining the spatial extent of objects. The simplest form of a geometric element for each dimension is called a simplex. For example, a node is a 0-simplex, an arc (a straight line consisting of two nodes) a 1-simplex, a triangle a 2-simplex, and a tetrahedron a 3-simplex.

Spatial position is defined by linking nodes to coordinates. Based on the concept of minimal objects and the notion that a minimal object in a higher dimension is composed of a specific number of minimal objects from lower dimensions, the following definitions can be given. (Note that some definitions in chapter 4 are repeated here for convenience.)

Definition 5.4-1: The metric dimension is defined by the number of linearly independent axes denoted by the coordinate tuple (Anton 1987).

For example, nodes are defined by coordinate pairs in 2-dimensional space, by (x, y, z) in 3-dimensional space, and by an n-tuple in n-dimensional space.

Definition 5.4-2: Any simplex of dimension n, called an n-simplex, is bounded by (n+1) geometrically independent simplices of dimension (n-1) (Amstrong, 1983; Egenhofer et al., 1989; Kinsey, 1993) and n+1 simplices of dimension 0 (which are in fact the vertices of K_{n+1} complete graph; see chapter 4; Finkbiner and Lindstrom, 1987).

For example, a tetrahedron (3-simplex, K_4 complete graph) is bounded by four triangles (2-simplices) and four nodes (0-simplices); a triangle (2-simplex, K_3 complete graph) is bounded by three arcs (edges of a triangle, 1-simplices) and three nodes; an arc (1-simplex, K_2 complete graph) is bounded by two nodes. Arcs are geometrically independent if they are not parallel and none of them is of length zero.

Definition 5.4-3: Confining analysis to an n-dimensional metric space, two n-simplices are always incident at a simplex of dimension n-1.

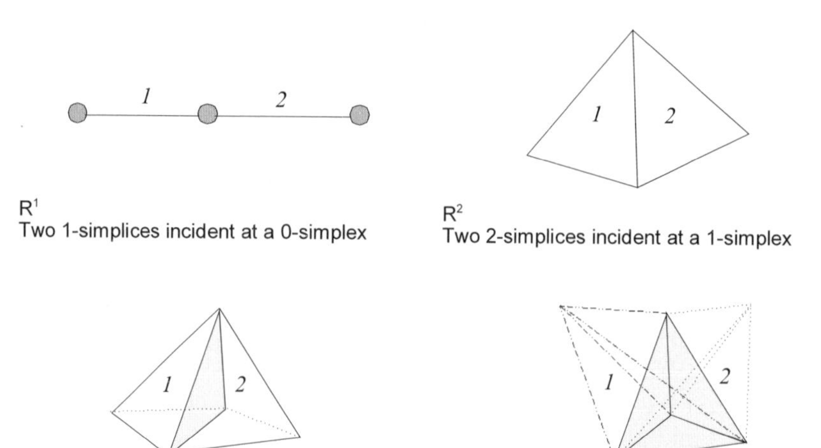

Fig. 5.5 Examples of two n-simplices incident at an (n-1)-simplex in R^n

The above definition can be turned into a component relation that is being shared. For example, in 1-dimensional space ($x \neq 0$), a node can be shared by at most two straight-line segments (whereas in two or higher dimensional space, a node can be shared by an infinite number of arcs); in a 2-dimensional space, an arc can be shared by only two triangles; in 3-dimensional space, a triangle can be shared by only two tetrahedrons. Similarly, in a 4-dimensional space, a tetrahedron can be shared by only two 4-simplices (see Figure 5.5 for the graphic illustration).

Note that the above definitions only hold for simplices; they do not hold for complexes.

Given the above three definitions, a generic n-dimensional data model can be derived following the logic we observed when extending our model from 2D to 3D. Figure 5.6 illustrates the nD data model. The generic data model can be illustrated elegantly, and it has the advantage that objects of dimensions higher than three need not be given names. The term 'simplicial network' is, therefore, introduced to refer to the nD network. The definition of a simplicial network can be given:

Definition 5.4-4: *An n-dimensional simplicial network is a network of simplices of different spatial dimensions, ranging from 0 to n-dimensions.*

Definition 5.4-5: *A finite set of simplices constitutes a complex that represents a spatial object.*

A simplicial network should also fulfil the generalized Euler characteristic described in section 5.6.

Let us recall the similarities between simplices and complete graphs mentioned in chapter 4 (see Figure 4.10). The definition of a simplicial network can be given in terms of graph theory.

Definition 5.4-6: A simplicial network is composed of a set of complete sub-graphs. The simplicial network itself need not be a complete graph. Either a simplicial network or each complete sub-graph can be, but not necessarily, a planar graph.

5.5 Single-theme and Multi-theme

The characteristic of a single-theme data model is that an instance of a feature type belongs to only one thematic class, and an instance of a geometric type (node, arc, triangle, tetrahedron) can be defined as a part of only one instance of a feature type (per theme). For a multi-theme data model, an instance of a feature type still belongs to only one thematic class, but an instance of a geometric type can be defined as a part of one or more instances of a feature type.

Within the multi-theme concept, two types of complexes must be distinguished. A homogeneous complex (feature) is a set of contiguous simplices of the same dimension, all relating to only one theme. A heterogeneous complex (overlapping part) is a set of contiguous simplices of the same dimension that relate to more than one theme. A heterogeneous complex is part of two or more homogeneous complexes. By introducing homogeneous and heterogeneous complexes, we can solve the problem of 'many-to-many' relationships between geometric primitives and features. The formal definition of a multi-theme integrated n-dimensional data model can thus be given.

Definition 5.4-7: A spatial object is represented by a complex. A complex is a finite set of simplices. Two or more complexes can overlap; their intersection yields a non-empty but closed and contiguous set of simplices that are embedded in the network structure.

Figure 5.6 shows the nD data model for the single-theme concept. Figure 5.7 shows the corresponding multi-theme data model.

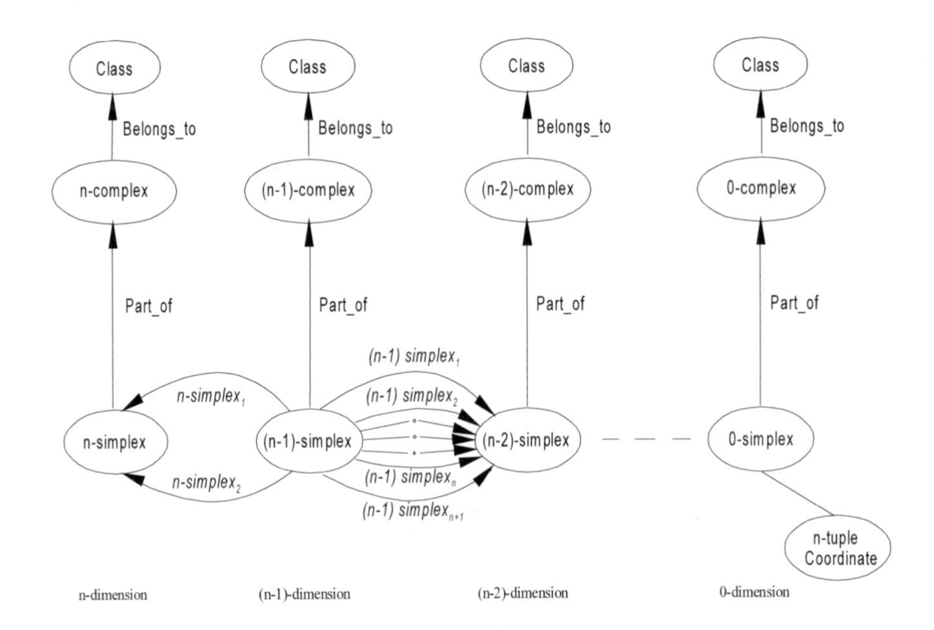

Fig. 5.6 A generalized n-dimensional data model for single-theme

The multi-theme data model can be seen as an extension of the single-theme data model, as it accepts objects that share the same spatial region. This extension means two or more objects can have overlapping parts (body, surface, line, point). A typical example is of layers of soil and a volume of ground water sharing the same spatial region.

5.6 Euler's Characteristics

This section presents the consistency aspect of the integrated data models with respect to the graph theory that is crucial for ensuring the integrity of a database structured by the simplicial network formation. The Euler characteristics described below can be used to design the consistency checking mechanism. General 2D-based GIS applied Euler's equality, which has been proven to work efficiently for planar graphs. In the case of simplicial networks, the TIN-based model still complies with Euler's equality, since it is limited to 2D topology. For the tetrahedral network, even though a 3-simplex (K_4) is a planar graph, its combination may yield a non-planar one. Moreover, the objects of dimension higher than 4 are clearly non-planar. Therefore, this section presents a more general solution that can apply to n-dimensions. The first part reviews Euler's equality for planar graphs as a basis. The second part introduces the generalized concept, the formalization and some proofs.

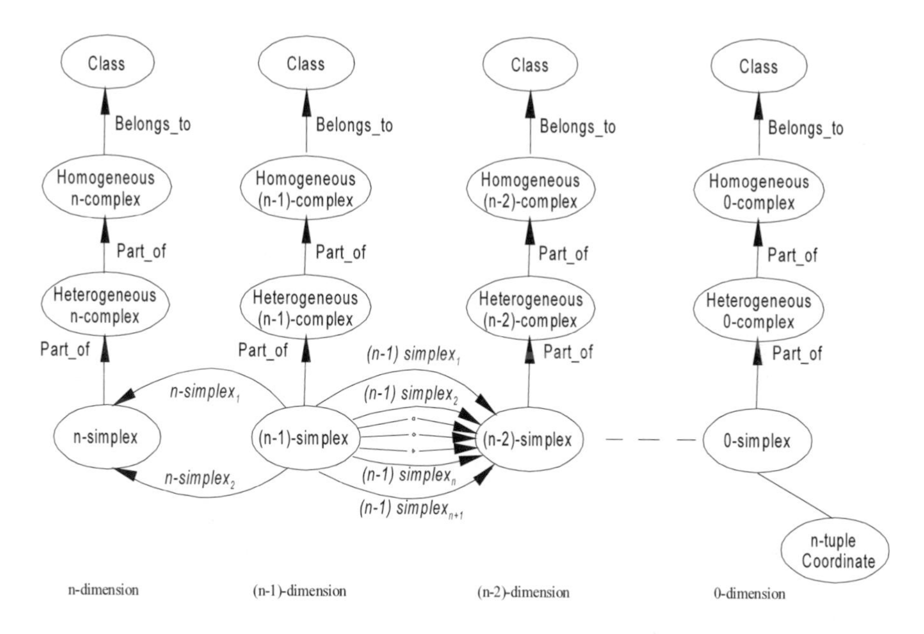

Fig. 5.7 Generic multi-theme data model for n dimensions

5.6.1 Euler's Equality

We recapitulate Euler's equality for a planar graph as described in chapter 4 by the following equation:

$$n + f = e + i$$

where,

n = number of nodes
f = number of faces
e = number of edges
i = degree of isolation

The degree of isolation indicates how many isolated regions are encountered. If the outer region is included in the graph, then it is also counted as a face; correspondingly, i should be increased by one. The above formula is applicable to the TIN-based data model.

5.6.2 The Generalized Euler Equality

To support the statement that simplicial networks of 3D and higher dimensions are non-planar, Kuratowski's theorem about the non-planarity of the graph is used.

Kuratowski's theorem: A graph is planar if and only if it contains no sub-graph homeomorphic to K_5 or $K_{3,3}$ (Kuratowski, 1930; see also chapter 4).

The above theorem implies that if a graph contains a sub-graph that is homeomorphic to K_5 or $K_{3,3}$, then it is a non-planar graph.

We recall from chapter 4 that two graphs are homeomorphic (equivalent) if and only if they are isomorphic, or both of them can be obtained from the same graph by inserting or deleting nodes of degree two (a node that has only two edges connecting to it). The degree of a node is defined by the number of edges that meet at that node.

K_5 is a complete graph (Figure 5.9) where $K_{3,3}$ is a complete bipartite graph. Recall again that a complete bipartite graph is a graph where the nodes are divided into two subsets (for example, a and b in Figure 5.8), such that each node in each subset is connected to every node of the other subset, one edge per pair of nodes.

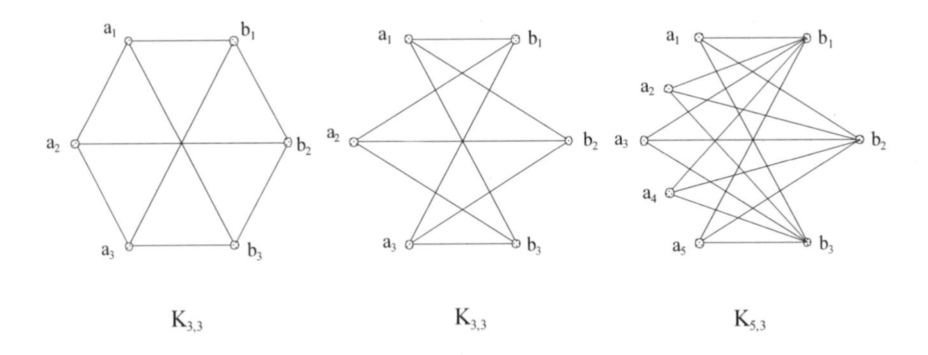

Fig. 5.8 Examples of complete bipartite graphs, $K_{a,b}$

By conducting a simple proof as graphically shown in Figure 5.9, it is clear that a tetrahedral network can contain sub-graphs isomorphic to K_5 or $K_{3,3}$.

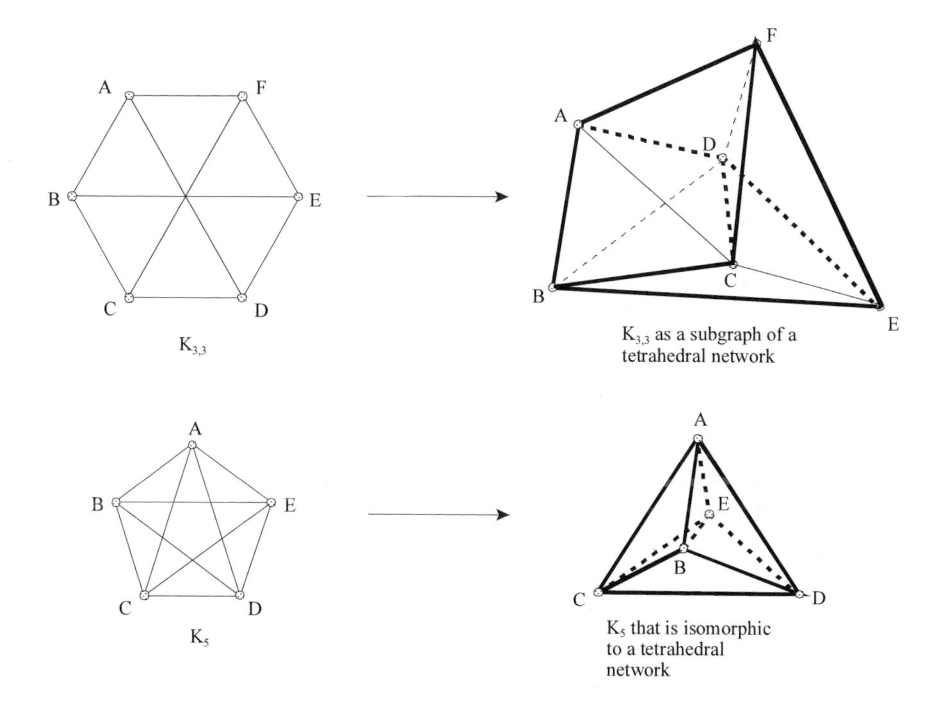

Fig. 5.9 Sub-graphs of a tetrahedral network that are isomorphic to K_5 or $K_{5,5}$. The thick lines indicates edges on the left side

The existence of such sub-graphs proves that a tetrahedral network is a non-planar graph and this also holds for any simplicial network of a higher dimension. The non-planarity implies that Euler's equality needs further generalization for it to be capable of application to the non-planar graph.

Sommerville (1929) has expressed an equation, similar to Euler's equality, for 3-cell complexes:

$$n - e + f - c = 1$$

where c = number of 3-cell complexes (see also Pigot, 1992). Pilouk et al., (1994) have presented the following equation applicable to a tetrahedral network:

$$Nodes + Triangles = Arcs + Tetrahedrons + 1$$

Note that the outer region is not included in the above formula. The variant for *n*-dimensions is:

$$0_{simplices} + 2_{simplices} + \ldots + k_{simplices} = 1_{simplices} + 3_{simplices} + \ldots + l_{simplices} + 1$$

where: k is even; $(0 \le k \le n)$
l is odd; $(1 \le l \le n)$
$0_{simplices}$ = number of nodes
$2_{simplices}$ = number of edges (arcs),
$k_{simplices}$ = number of simplices of dimension k
$l_{simplices}$ = number of simplices of dimension l

The above equation can be used to verify whether the simplicial network is well constructed. The imbalance indicates that the simplicial network is ill-formed, that is to say, having either free points, or intersecting edges, or faces presented in the network (Figure 5.10).

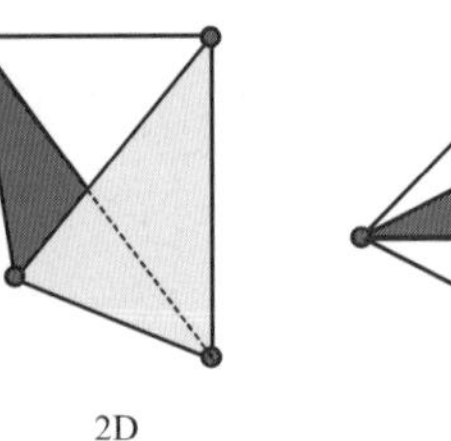
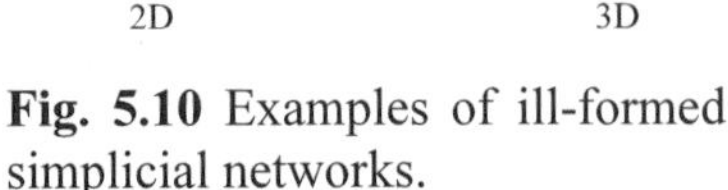

Fig. 5.10 Examples of ill-formed simplicial networks.

Another variant of the generalized Euler equality for an n-dimensional complex is:

$$0_{complexes} + 2_{complexes} + \ldots + k_{complexes} = 1_{complexes} + 3_{comolexes} + \ldots + l_{complexes} + i$$

where:

k is even; $(0 \le k \le n)$
l is odd; $(1 \le l \le n)$
i = degree of isolation

It is important to note that the degree of isolation must be determined correctly. There are different kinds of degree of isolation indicating the number of isolated objects. The isolated objects to be determined are:

- nodes with no connection to any arcs,
- arcs (a dangling arc does not fall into this type),
- faces (a dangling face does not fall into this type),
- bodies, for example, holes in a body.

For n-dimensions, isolated objects of dimension 4 and above should also be included in i. Nevertheless, the type of isolation must be specified as a convention for each data model, so that an imbalance indicates that the convention is not met, and the system can issue a warning to the user. In the case of the integrated data model based on the simplicial network, the degree of isolation is equal to 1, because no isolated object other than the network itself is allowed.

The above formulae can be rewritten in a general form:

for one simplicial network with no isolation:

$$\sum_{k=0}^{k\leq n} (-1)^k N_k = 1$$

for a simplicial network with *I* degree of isolation:

$$\sum_{k=0}^{k\leq n} (-1)^k N_k = \sum_{k=0}^{k\leq n} I_k$$

where,
n = dimension number
N_k = number of k-simplices
I_k = number of isolated objects or sets of mutually connected objects.

A mathematical proof of the above equation is given in Appendix A.

5.7 Discussion

The developed simplicial network data model is based on four important concepts:

- formal data structure
- a constrained Delaunay network
- simplicial complexes
- graph theory.

FDS helps define representations of real world objects with respect to their relationships with geometric and thematic components. A constrained Delaunay network provides the basic concept for representing spatial units suitable for computation where existing knowledge represented in a form similar to FDS can be considered to be the constraints. So, the computation result can be adapted to the situation in reality. Simplicial complexes and graph theory provide sound mathematical foundations for the simplicial network data model and rigorously support the generalization of this concept.

Both the direct and indirect representation of real world objects can be accommodated by the simplicial network data model. It permits refinement of the knowledge about the reality by deriving new information from existing facts presented as the direct representation type. The locality property of a simplicial network permits the adding of new facts into the spatial model without undue disturbance of the model as a whole. The local property applies to the elimination or updating of components of the model that no longer represent reality well. The adaptability, a property of an irregular network, makes modelling the variation aspect of the reality possible. Since the model is based on the complete tessellation of space, there are various means of navigation within the model, for example, using metric computation, order, or topology. With respect to the volume needed for storage of this kind of spatial model, the amount of data is expected to be less than that needed to store the components of a 3D spatial model separately. For example, storing terrain relief and terrain features in two separate data sets implies storing redundant elements where two representations coincide. The finiteness and convexity properties of each element of the model help simplify many operations. The data model complies with the generalized Euler characteristics, which can be used for checking logical consistency of the model with respect to its geometrical aspect. This consistency checking and some of its examples will be discussed in chapter 8.

Chapter 6 THE LOGICAL DESIGN

With the conceptual design of the integrated data model (IDM) presented in the previous chapter, we proceed to the logical design stage aiming at the unified data structure (UNS). This chapter explains the translation of the IDM into two kinds of UNS, using the relational and object-oriented approaches respectively. In contrast with the conceptual model, which is independent of the type of system and computing platform, the data structure comes closer to the implementation stage. The type of database management system (DBMS), which depends on a hierarchical, network, relational or object-oriented concept, has to be selected. The object-oriented approach contains the concepts of network and hierarchy and so demands more implementation effort. Not only do all objects, but also the methods of accessing each object, need to be carefully defined. Each DBMS type may only be available on one specific computing platform. The advantages and disadvantages in terms of speed and efficiency, ease of implementation, system maintenance and upgrade, and compatibility, have to be weighed to select an appropriate system for the implementation.

Since the purpose of this chapter is to describe the approach to translating the IDM presented in chapter 5 into UNSs, only a few single-theme variants of the IDM have been selected as examples. The same approach can be followed for the other variants.

6.1 Relational Approach

The reasons for using the relational approach include:

- ease of implementation; users can concentrate on the application rather than concern themselves about data access, since this is taken care of internally by the DBMS.
- flexibility; the data structure can be readily extended or modified to delete or add more attribute columns, change the number of characters in a string data type, and so forth.
- availability of various database management systems (Oracle, Informix, dBASE, Interbase) on different computing platforms and operating systems (PC with DOS, Windows, or UNIX; workstation, mini or mainframe with UNIX, VMS, or Windows).
- availability of software libraries and APIs (ODBC) and query languages (SQL, QBE).

- possibility of importing and exporting data to other systems, such as a spread sheet, or a word processor.

An important reason for choosing the relational approach is its maturity in providing a rigorous procedure for mapping a data model to a data structure. This process is known as normalization (see chapter 4). It is the mechanism ensuring the data integrity of the database in the face of updating anomalies. We obtain a set of skeleton tables here, using Smith's normalization procedure as presented in Roessel (1986), Kufoniyi (1989), Bouloucos et al., (1990), Ayugi (1992), Pilouk and Tempfli (1992), Chhatkuli (1993). The relational approach also helps clarify and create the understanding that spatial relationships called for when establishing object-oriented data structure in which the same kinds of relationships have to be represented.

6.1.1 Relational Data Structure for TIN-based Model

Following the five steps of Smith's normalization described in chapter 4, a TIN-based relational data structure is constructed as follows:

6.1.1.1 Constructing Dependency Statements

This step starts with the identification of the data fields to be stored in the database. In the data model in Figure 6.1, data fields are encompassed by ellipses, and the relationships are the labels on the lines connecting pairs of ellipses. The relationship between each pair of fields is analysed and then translated into a dependency statement. The list of dependency statements is given below (Pilouk and Tempfli, 1993).

1. A surface feature, which is identified by a *SID*, belongs to one *SCLASS* surface feature class.
2. A line feature, which is identified by a *LID*, belongs to one *LCLASS* line feature class.
3. A point feature, which is identified by a *PID*, belongs to one *PCLASS* point feature class and is represented by one *PNODE* node number.
4. Each *NODENR* node has a position given by one *X* x-coordinate, one *Y* y-coordinate, and one *Z* z-coordinate.
5. An arc is identified by *ARCNR;* it has one *Node1* starting node and one *Node2* ending node, and at most one *Tri1* triangle on its left side and at most one *Tri2* triangle on its right side.
6. An *ARCNR* arc represents at most one *ALID* line feature.
7. A triangle is identified by *TRINR* and represents at most one *TSID* surface feature.

6.1.1.2 Mapping from Dependency Statements into Dependency Diagram

From the above list of dependency statements, the corresponding dependency diagram can be drawn as in Figure 6.1. The attributes (data fields) are shown within bubbles. A line between two bubbles indicates a relationship between one data field and the other. A single-headed arrow indicates that it is a single-valued dependency; a double-headed arrow indicates a multi-valued dependency. More than 1 bubble covering a data field indicates that not all the relationships may apply to every instance of the data field. For example, an *ARCNR* should have a left and a right triangle (*tri1* and *tri2* respectively) but may not be part of a line feature. A number adjacent to a line between two bubbles indicates the dependency statement number. The indicator of the number of differently named fields having a common field type (*eg TRINR*, *tri1*, and *tri2* are of the same field type representing triangle identifiers) is the *domain flag;* it is shown as a number in a small triangle (see Figure 6.2).

6.1.1.3 Composing Relational Tables from Dependency Diagram

Tables are first composed from the single-valued dependencies and then from the multi-valued dependencies. A bubble with no arrow pointing to it becomes a primary key field in one table. A target bubble becomes a data field in the same table. A bubble pointed to by an arrow and having a domain flag also becomes a foreign key field in the same table. In the case of multi-valued dependency, all the data fields with emanating arrows comprise primary keys. Special care should be taken here if there are more than three fields comprising a primary key; the table may not be practicable, since it would result in bad response times. The solution is to split the table into two by introducing a surrogate key acting as the primary key in one table and as a foreign key in the other. The following tables result (see also Figure 6.1):

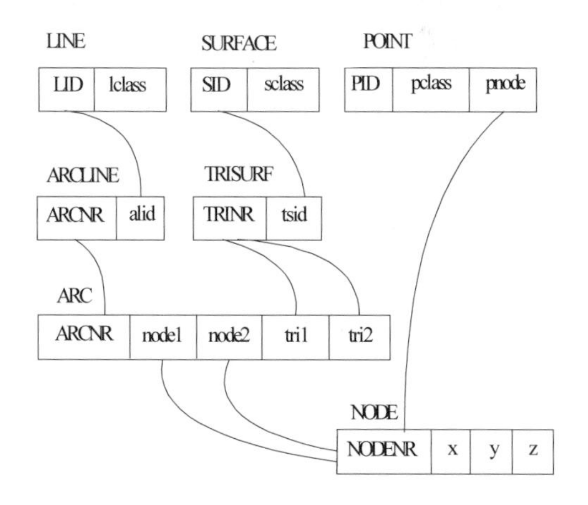

Fig. 6.1 TIN-based relational data structure

R1: NODE (NODENR, x, y, z)
R2: ARC (ARCNR, node1, node2, tri1, tri2)
R3: TRISURF (TRINR, tsid)
R4: ARCLINE (ARCNR, alid)
R5: POINT (PID, pclass, pnode)
R6: LINE (LID, lclass)
R7: SURFACE (SID, sclass)

For convenience, the relational tables are labelled here by codes R1 to R7. Each table has a table name shown outside the bracket. Inside the bracket is the primary key, with its name shown in capital letters, and the set of attributes. The tables R1 and R2 represent geometric primitives and all the necessary topological relationships; for example, an arc has two nodes for its start and it end end, and two triangles on the left and the right side. R3 and R4 represent part-of relationships between geometric primitives and features; they are the same as R5, except that R5 also represents the thematic classification resulting from the one-to-one relationship between a node and a point feature. R6 and R7 represent thematic classes for line and area features.

6.1.2 Relational Data Structure for a TEN-based Model

The TEN-based data model can be mapped into a relational data structure by following the same procedure as for a TIN-based model. Most of the dependency statements are the same as for the TIN-based model. Some statements, however, have to be modified and some additional statements are required.

The following dependency statements yield the following:

1. A body feature, which is identified by a *BID*, belongs to one *BCLASS* body feature class.
2. A surface feature, which is identified by a *SID*, belongs to one *SCLASS* surface feature class.
3. A line feature, which is identified by a *LID*, belongs to one *LCLASS* line feature class.
4. A point feature, which is identified by a *PID*, belongs to one *PCLASS* point feature class and is represented by one *PNODE* node number.
5. An arc is identified by *ARCNR* and has one *NODE1* starting node and one *NODE2* ending node.
6. Each *NODENR* node has a position given by a one *X* x-coordinate, one *Y* y-coordinate, and one *Z* z-coordinate.

7. A triangle is identified by *TRINR* and represents at most one *TSID* surface feature; it has at most two tetrahedrons *TET1* and *TET2* attached to it, one on each side of the facet. It has at most three edges, *EDGE1*, *EDGE2* and *EDGE3*.
8. An *ARCNR* arc represents at most one *ALID* line feature.
9. A tetrahedron is identified by *TETNR* and represents at most one *TBID* body feature.

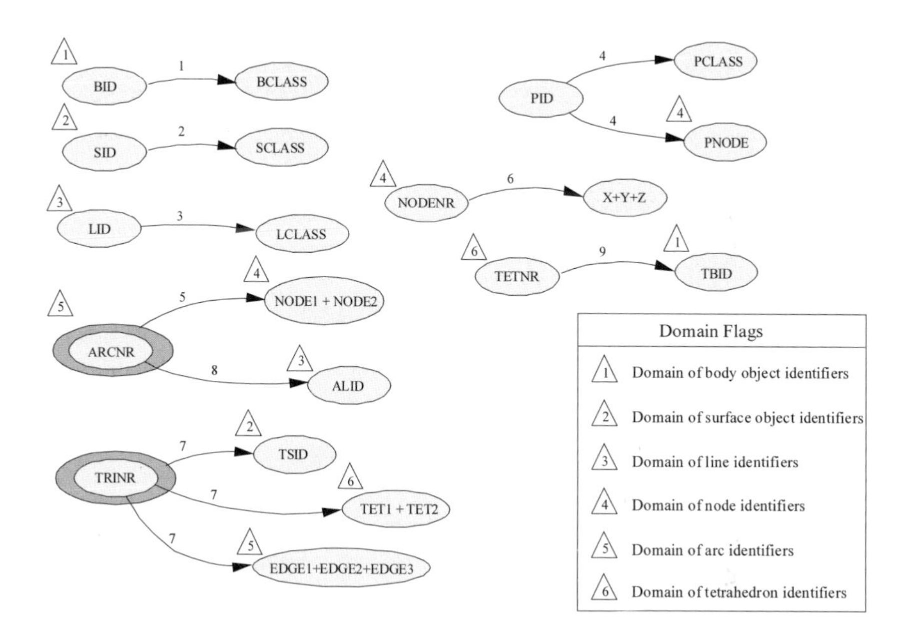

Fig. 6.2 Dependency diagram of the tetrahedron-based data model

Figure 6.3 shows the dependency diagram derived from the above list of dependency statements.

The following ten relations (tables) are obtained from the normalization process.

R1: Node (NodeNr, x, y, z)
R2: Arc (ArcNr, node1, node2)
R3: Triangle (TriNr, tet1, tet2, edge1, edge2, edge3)
R4: Tetra (TetNr, tbid)
R5: TriSurf (TriNr, tsid)
R6: ArcLine (ArcNr, alid)
R7: Point (Pid, pclass, pnode)
R8: Line (Lid, lclass)

R9: Surface (Sid, sclass)
R10: Body (Bid, bclass)

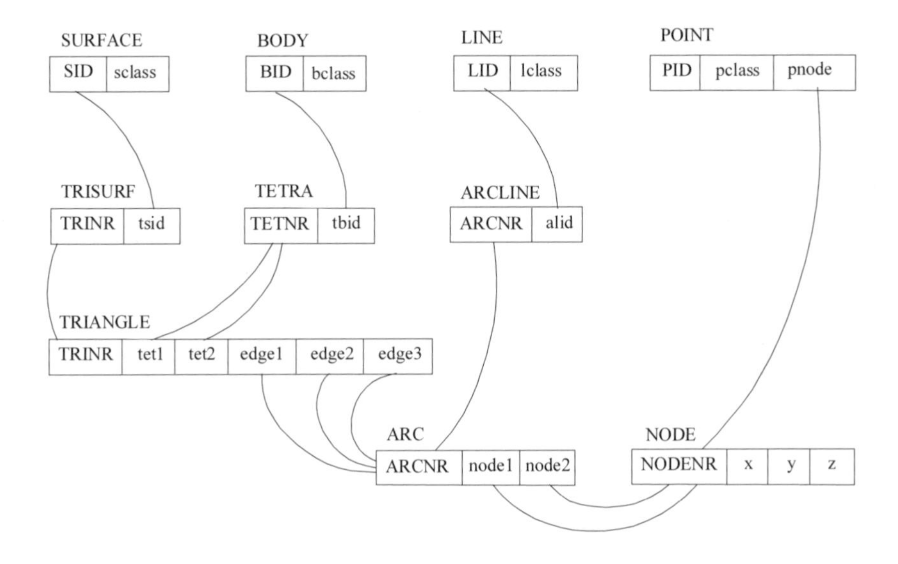

Fig. 6.3 A TEN-based relational data structure

R1, R2 and R3 can be regarded as geometry tables. R4, R5 and R6 are geometry-feature tables. R7 is a geometry-feature-class table. R8, R9 and R10 are class tables. Note that table R4 maintains no other information than a tetrahedron number (*TetNr*) and an identifier of the body feature (*TBID*) to which it belongs. We can only search for the geometric components of a tetrahedron of interest via the R3 (Triangle) table by matching the attribute value of the tetrahedron (*TetNr*) with either attribute value of *tet1* or *tet2*. Once the match is found, the next step is to get each of the three attribute values of *edge1*, *edge2* and *edge3* of the R3 table as a key to search for the match with the *ArcNr* in the R2 (Arc) table. If the match is found, we must get the attribute values of *node1* and *node2* and use each of them to search for the match with the *NodeNr* in the R1 (Node) table to get the coordinates x, y, and z for each respective node. In this way, we can use this database for 3D interpolation and for responding to a wide range of queries.

6.1.3 Relational Data Structure for an n-dimensional Data Model

Since the generic integrated n-dimensional data model has been presented in the previous chapter, mapping into the corresponding n-dimensional UNS will not be further elaborated here. The procedure followed in the TIN-based model can also be applied to a TEN-based data model. This section is restricted to the end results, included here for the sake of completion.

By observing the number of tables in the TIN-based and TEN-based UNSs and applying mathematical induction, it is possible to intuitively predict the number of tables necessary for the n-dimensional UNS. There are *n* geometric tables, *n* geometry-feature tables, *1* geometry-feature-class table (i.e., for the link between node and point feature and class), and *n* feature-class tables. The set of relational tables will resemble the following:

Geometry

G_1 (S_0Nr, Crd_1, Crd_2, Crd_3, ..., Crd_n);
G_2 (S_1Nr, S_{01}, S_{02});
G_3 (S_2Nr, S_{11}, S_{12}, S_{13})
G_4 (S_3Nr, S_{21}, S_{22}, S_{23}, S_{24})
...
G_{n-1} ($S_{(n-2)}$Nr, $S_{(n-3)1}$, $S_{(n-3)2}$, $S_{(n-3)3}$, ..., $S_{(n-3)(n-1)}$)
G_n ($S_{(n-1)}$Nr, $S_{(n-2)1}$, $S_{(n-2)2}$, $S_{(n-2)3}$, ..., $S_{(n-2)n}$, S_{n1}, S_{n2})

Geometry-feature

GF_1 (S_1Nr, $c_1$id)
GF_2 (S_2Nr, $c_2$id)
GF_3 (S_3Nr, $c_3$id)
GF_4 (S_4Nr, $c_4$id)
...
GFn (S_nNr, $c_n$id)

Geometry-feature-class

GFC (C_0id, C_0class, S_0Nr)

Feature-class

FC_1 (C_1id, C_1class)
FC_2 (C_2id, C_2class)
FC_3 (C_3id, C_3class)
FC_4 (C_4id, C_4class)
...
FC_n ($C_n$id, $C_n$class)

where Crd_i is a coordinate component,

S_i represents the simplex or geometric primitive of i dimension,

C_i represents the complex or feature of i dimension.

S_{ij} represents the number j of i-simplex

C_{ij} represents the number j of i-complex

6.2 Object-oriented Approach

Although a relational database approach yields several advantages, certain important aspects are still lacking, which the object-oriented approach promises to fulfil. These aspects are:

- the relational approach is based on the Cartesian product. The joint operation on several tables causes a long response time, particularly for large amounts of data commonly found in GIS. The object-oriented approach includes the hierarchical and network data structures that can efficiently represent topology and facilitate navigation among different elements in the database, and so is likely to have a better response time.
- a more complete and precise control over each individual object, especially where considerable ambiguity exists, as may happen when there are many different types of objects stored within the same database.
- re-usability and extendibility of database management API (Application Program Interface) with no modification to the source code. These requirements mostly come from the community of developers, where the source code needs to be protected and hidden from users for commercial reasons. The API is implemented and compiled into a computer object code that encapsulates the objects, their attributes and accessing methods together in a form similar to software libraries, where only function names, methods of calling and function parameters are provided. The reusability and extendibility of the API are provided through the inheritance mechanism. Users can further develop the API to fit their requirements by deriving new classes from existing classes, using an object-oriented compiler, such as C++, Smalltalk, or Object Pascal.

This section reports a study applying the object-oriented concept to the structured geoinformation based on the integrated data model. The focus is on the definition of objects and the design of object class hierarchies.

An object-oriented approach provides many alternatives to the design and implementation with respect to different abstractions of the real world. Worboys et al., (1990), Kainz and Shahriari (1993) have presented similar designs in which the thematic class is defined as a parent that passes all aspects onto the geometric class. Their approach may be considered too rigid if multiple representations are needed. Multiple representations require different types of geometry to be chosen for the representation of an object, depending on the level of abstraction. If the geometric representation for an object has already been fixed at the design stage, it would not be possible for the user to select any other kind of geometric representation. If for example it is decided to represent a road as a line-feature, it would not be possible to represent the road later as a band, as might be needed for abstraction on a larger scale, since the band is an area-feature. The whole hierarchy would have to be redesigned for every different level of abstraction, which could result in many classes. This approach may therefore be regarded as an *ad hoc* solution.

The selected approach follows as strictly as possible the conceptual model defined by (Molenaar, 1989). This conceptual model offers a natural way of handling geoinformation, especially when considering the aspect of object creation that relates to data acquisition. This approach does not fix the geometric representation of a feature at the design stage. It divides the components of a feature into two hierarchies. The inheritance hierarchy is used for the thematic attributes and the aggregation hierarchy for the geometric attributes. These two hierarchies are only combined at runtime (the construction phase of the spatial model), thus allowing the user to select different types of geometric representation for a feature. The requirement for multiple representations can thus be fulfilled. Comparing the two approaches, the latter is more versatile, but it is more difficult to implement and requires highly skilled software engineering.

In the following sections, the translation of the IDM into object definitions follows, using the abstraction mechanisms of the object-oriented paradigm presented in chapter 4, namely classification, specialization, aggregation, and inheritance. The top-down approach starts from the most generalized class and proceeds to the most specialized class. Part of the implementation to UNS is illustrated using the C++ object-oriented programming language.

6.2.1 Object-oriented Definition of a Spatial Object

Recall that the abstraction of real world objects consists of two parts: geometric and thematic (see Figure 6.4). The geometric component contains information about shape, location and topology, while the thematic component contains human knowledge about other properties of the object (colour, name, ownership, function, and so forth). When applying a concept using the relational approach, the geometric and thematic attributes are linked through a feature identifier (see section 6.1.1). In an object-oriented approach, the geometric and thematic components are realized as objects that can be tied together by a 'feature object' through the encapsulation mechanism.

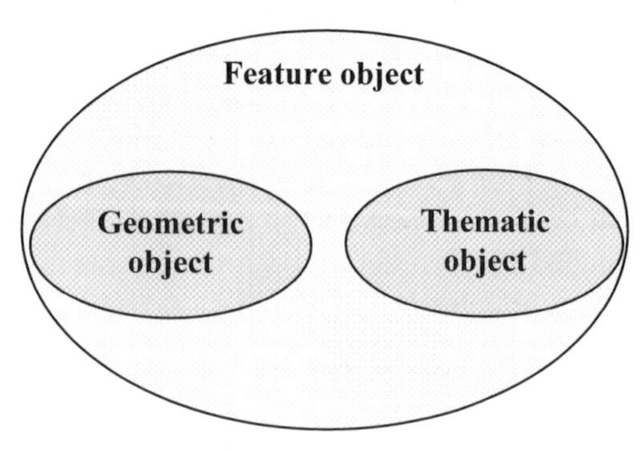

Figure 6.4 Object-oriented definition of a real world object. Feature object encapsulates geometric component and thematic component.

Since the object-oriented approach uses terminology in a similar way to our normal descriptions of reality in the conceptual design stage, it is important to note that the discussion in this section is limited to the logical design stage in object-oriented programming terminology. The term 'class' is an abstract data type (ADT), whereas the term 'object' is used to refer to an instance of a class in object-oriented programming terminology. The terms class and ADT are used interchangeably.

6.2.2 Object-oriented Design Based on IDM

Classification

Molenaar (1993) provides a rationale for arriving at an object-oriented design. We are concerned with the logical design of an object-oriented data structure which defines a scheme for the storage of information about the spatial object. The representation of a real world object can be translated into three ADTs; ADTFeature, ADTGeometry, and ADTTheme. The ADTFeature class at the top level of the hierarchy aggregates the ADTGeometry and ADTTheme classes at the lower level. The ADTFeature class defines the storage of a representation of a real world object as a whole. The storage of geometric and thematic descriptions about the real world object are defined by the ADTGeometry and ADTTheme classes respectively. A feature is an instance of an ADTFeature class. Likewise, a geometric object is an instance of an ADTGeometry class and a thematic object is an instance of an ADTTheme class.

Each of the three ADTs is considered as a general class with its own hierarchy and which still has to be defined further. A general class is subdivided into more specific classes to any desired level of refinement. The common status and behaviour of subclasses characterize their general class. Given the three general classes, we have to deal with three class hierarchies which have to be related at an appropriate time based upon the user's requirements and context (Molenaar 1993). Among the three general classes, the ADTFeature class has a 'primus inter pares' position. It must be considered to be at a higher level, since it constitutes and defines an aggregation hierarchy from the other two classes.

Thematic class

The ADTTheme class defines a data structure for the storage of thematic information which is highly related to the application domain in geoinformation, such as land use, transportation networks, water bodies, and the like. This class consists of information about common attributes and behaviours of descendant thematic objects. The purpose of having this class is to facilitate the definition of an inheritance hierarchy, minimizing redundancies and allowing re-usability between thematic information.

There follows a basic ADT for the storage of thematic information.

class	: ADTTheme
description	: general representation of thematic components of a spatial object
parent	: none
attributes	: code, name, texture, colour, ...
methods	: create, delete, show code, modify code, show name, modify name, ...
constraints	: ...

Geometric class

The geometric description of a spatial object is stored and maintained in the ADTGeometry class. This class defines a hierarchy of geometric primitives which comprise the geometric descriptions of a spatial object. The class provides a general data structure for the storage of components, describing the shape of each feature, its georeferencing scheme and its topological relationships with other features. One important aspect is that every geometric object has to be referred to its ADTFeature class. This relates to the everyday life situation, where subordinates should always know their superiors. An ADTFeature object is akin to the boss who represents and rules the group of subordinates. This assumption helps us make the organization more natural and efficient.

A basic ADTGeometry class is defined as follows:

class	: ADTGeometry
description	: general representation of geometric components of a spatial object
parent	: none
attributes	: identifier, reference to a feature class object (part of), ...
methods	: create, delete, show identifier, modify identifier, show feature, modify feature, display graphics, ...
constraints	: on creation, requires input of object identifier and reference to a feature object from the creator; on creation, sends a request to the reference feature object to update the geometric container of the feature, ...

Feature class

The ADTFeature class plays a central role in the representation of the real world. This class provides the interface between the users and the system. The class is also the entry point for the user to retrieve or store all components of the feature. The ADTFeature class is an aggregate class. Any instance of this class is a composite object, consisting of two components; an ADTGeometry object, and an ADTTheme object. In other words, both ADTGeometry and ADTTheme objects form 'part-of' an ADTFeature object.

Class	: ADTFeature
Description	: general representation of a spatial object
Parent	: none
Attributes	: identifier (to interface with users), reference to ADTTheme object, reference to the collection of ADTGeometry objects, ...
methods	: create, delete, add to geometric container a reference to ADTGeometry object, delete from geometric container a reference to ADTGeometry object, (graphics) display geometry, display thematic properties, ...
constraints	: on creation, requires identifier and reference to ADTTheme object from the creator, ...

6.2.3 Specialization of Classes

As we have said, each of the above classes can be further refined as more detailed objects. The following sections show the construction of class hierarchies using the specialization mechanism, resulting in inheritance hierarchies.

6.2.3.1 Thematic Hierarchy

The class ADTTheme can be specialized as various subclasses, such as road, railway, river, control point. The construction of this hierarchy is very subjective, depending on the user's point of view and application. There are, however, many advantages in modelling thematic information using an object-oriented approach, especially when an object has to be related to several themes at the same time. The object-oriented approach provides a straightforward solution to this representation through the multiple inheritance mechanism. A class can inherit properties from more than one parent class. Such a class then represents a combination of themes.

An example, taken from Figure 6.6, is the TRiver class which can be seen as 'is-a' TWaterBody and TNatural transportation network at the same time. By aggregating an object that belongs to the thematic hierarchy (for example, class TRiver) as the component of an object that belongs to the feature hierarchy (for example, line feature class), the object belonging to the feature hierarchy automatically carries multiple thematic information. However, ambiguity may arise in such a case; two parent classes may have attributes or methods of the same name and so need special attention and appropriate resolution. The designer of this class must decide from which parent the new class inherits the properties, otherwise the ambiguous properties have to be completely overridden. It may be necessary to resolve the ambiguity by setting up consistency rules as detailed by Kufoniyi (1995). Egenhofer and Frank (1989) and Kainz and Shahriari (1993) have reported some other examples emphasizing the construction of the thematic hierarchy. The specialization of thematic class, however, is not elaborated further here.

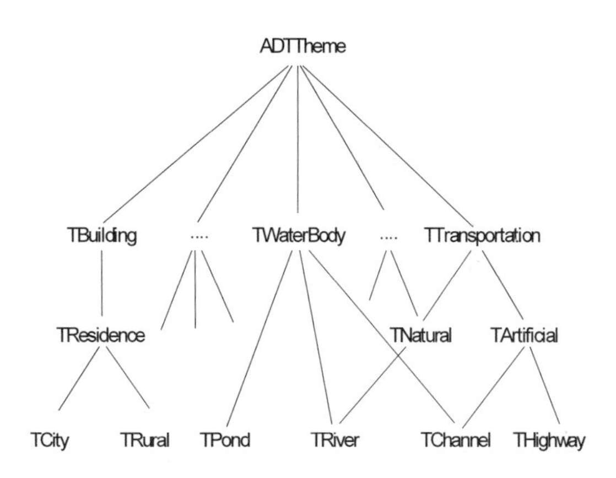

Fig. 6.5 Thematic class hierarchy (adapted from Egenhofer et al 1989)

6.2.3.2 Geometric hierarchy

The class ADTGeometry can be specialized for each geometric primitive or simplex—node, arc, triangle, tetrahedron and so forth, as shown in Figure 6.6. These classes inherit properties from the parent class ADTGeometry; each of them contains only the additional status and behaviours that are different from its ancestor.

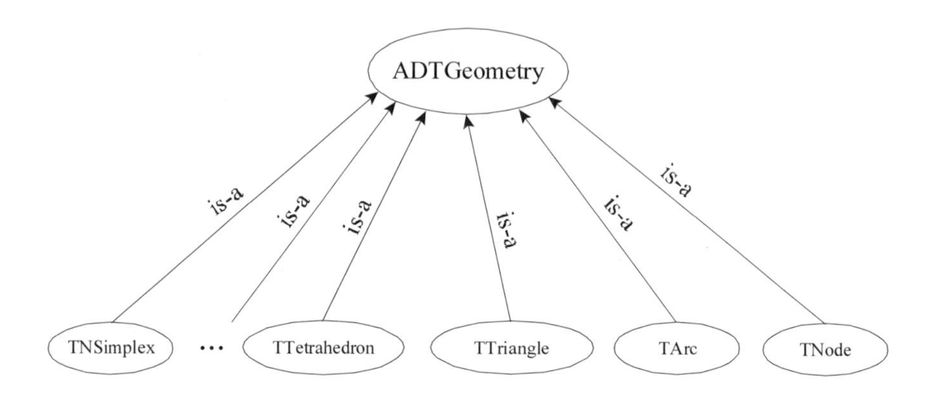

Fig. 6.6 Geometric class hierarchy.

The following shows the class descriptions of the features.

class	: TNode
description	: 0D geometric component of the representation of a spatial object
parent	: ADTGeometry
attributes	: x, y, z coordinate, perspective transformed coordinate (xp,yp)
methods	: get coordinates, modify coordinates, (graphics) display, 2D, 3D, perspective transformation, ...
constraints	: requires 3D coordinates, reference to ADTFeature object, and a geometric identifier from the creator

class	: TArc
description	: 1D geometric component of representation of a spatial object
parent	: ADTGeometry
attributes	: references to two TNode objects as begin and end nodes, references to two TTriangle objects on its left and right sides.

methods	: create, delete, (graphics) display 2D, 3D, ...
constraints	: requires identifier, references to two TNodes and ADTFeature objects from creator, ...

class	: TTriangle
description	: 2D geometric component of representation of a spatial object
parent	: ADTGeometry
attributes	: references to three TNode objects as its vertices, three TArc objects as its edges, three TTriangle objects as neighbour triangles, slope, parameters of plane in normal form (a, b, c, d), ...
methods	: create, delete, get edges, get neighbours, get slope, get plane parameters, interpolate elevation for a given x,y coordinate, interpolate locations for a given z coordinate, (graphics) display 2D, display 3D, shade, ...
constraints	: requires a geometric identifier, references to three TArc objects, and reference to an ADTFeature object, ...

class	: TTetrahedron
description	: 3D geometric component of representation of a spatial object
parent	: ADTGeometry
attributes	: references to four TTriangle objects as its faces, four TArc objects as its edges, four TNode objects as its vertices, four TTetrahedron objects as its neighbours, ...
methods	: create, delete, get vertices, get edges, get faces, get neighbours, interpolate value for a given x,y,z coordinates, interpolate contour surface, (graphics) display 3D, shade, ...
constraints	: requires a geometric identifier, references to four TNode objects, and reference to an ADTFeature object, ...

6.2.3.3 Feature Hierarchy

The class ADTFeature is specialized into four specific classes: point, line, area and body, as shown in Figure 6.7. Each derived class has its specific behaviours and attributes in addition to the behaviours and attributes of the parent class ADTFeature. A simple example is the draw operation. Drawing a point may only require drawing a pixel on a screen, while drawing a line, or an area, requires additional operations. The topology has to be used to navigate in the database to obtain all the nodes and their links before the pixels can be drawn along the line, or along the boundary of the area. The specialization also helps streamline the handling of the geometry and topology of each particular subclass. The design of related functions can be concentrated on specifically for each one in turn, with no fear of their interfering with each other, even if the functions of the different objects have the same function names.

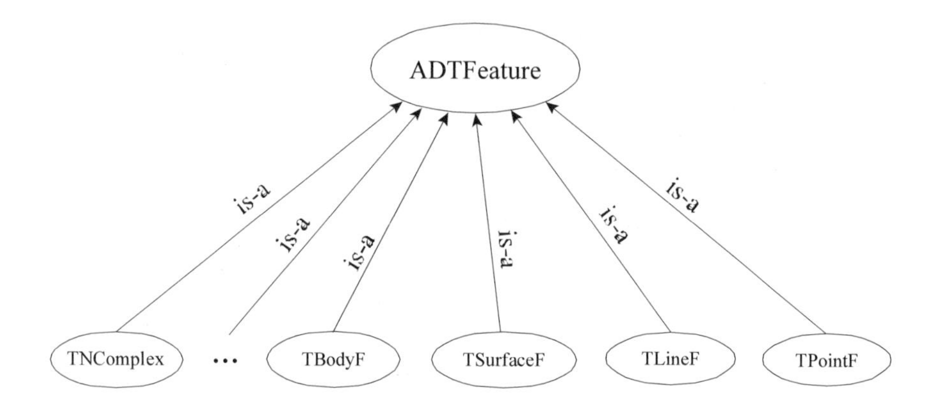

Fig. 6.7 Feature class hierarchy

The following shows the descriptions of the features:

class : TPointF
description : 0D representation of a spatial object
parent : ADTFeature
attributes : ...
methods : create, delete, display geometric (draw node, 2D, 3D), display thematic, ...
constraints : on creation, requires an identifier, references to ADTGeometry and ADTTheme objects from the creator, ...

class : TLineF
description : 1D representation of a spatial object
parent : ADTFeature
attributes : bounding rectangle, ...
methods : create, delete, display geometric (draw all component arcs as 2D, 3D), display thematic, ...
constraints : on creation, requires an identifier, references to ADTGeometry and ADTTheme objects from the creator, ...

class	: TSurfaceF
description	: 2D representation of a spatial object
parent	: ADTFeature
attributes	: bounding rectangle (cube)
methods	: create, delete, display geometric (draw all component triangles as 2D, 3D, shade), display thematic, ...
constraints	: on creation, requires an identifier, references to ADTGeometry and ADTTheme objects from the creator, ...

class	: TBodyF
description	: 3D representation of a spatial object
parent	: ADTFeature
attributes	: bounding box (cube)
methods	: create, delete, display geometric (draw all component tetrahedron as, 3D, shade), display thematic, ...
constraints	: on creation, requires an identifier, references to ADTGeometry and ADTTheme objects from the creator, ...

6.2.4 Aggregation of Objects

The ADTFeature class forms an aggregation hierarchy by taking objects belonging to the geometric and thematic hierarchies as its components (see Figure 6.8). This is a stage of assembling or manufacturing an instance of the ADTFeature class. Subclasses of this class, for example, TPointF, TLineF, TSurfaceF, TBodyF, are also of aggregate types; a TLineF object may consist of many TArc objects. For each subclass of the ADTFeature, the actual aggregation has to be done at runtime. This is because it is not possible to know at the design phase which specific class in the thematic hierarchy will be its thematic component. The dynamic referencing mechanism is the solution to this problem. The technique is first to define the aggregation, using the reference to a generic class (ADTTheme). During runtime, the user selects the more specific class (for example, class TRoad). Dynamic inheritance and aggregation take place here. The class that aggregates the class TRoad into the class TLineF is, in fact, derived at runtime. The TLineF object knows at that moment that its thematic component is of the specific class TRoad, which is the descendant instead of the generic class ADTTheme. The reference to class ADTTheme is then changed to class TRoad.

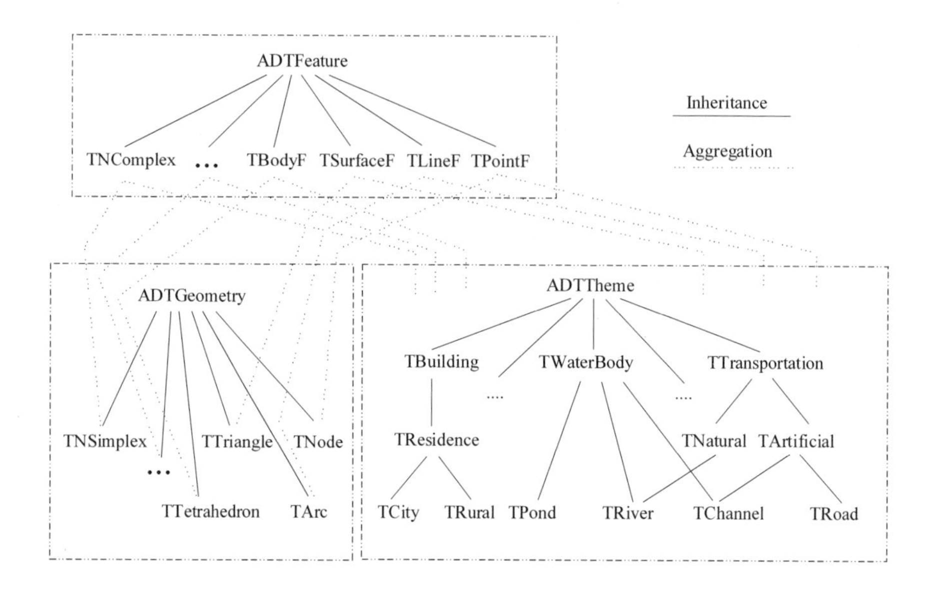

Fig. 6.8 Relationship between class hierarchies

6.2.5 Creation of Objects

In addition to the classes defined above, the system must provide container classes, each of which is specific to the objects of the ADTTheme, ADTGeometry and ADTFeature classes. The objects for each class should be created in an appropriate sequence. ADTTheme objects are the first to be created and registered into the container of ADTTheme. In practice, users should first define their own thematic hierarchies according to the purpose of the application. For example, if the geoinformation is to serve the management of a road network, the thematic hierarchy should start from 'general road' and then specialize down to 'primary road', 'secondary road', 'highway', 'superhighway', and so on.

The ADTFeature objects are created next. Every instance of this class must be registered into the container of ADTFeature. The user defines which theme is to be represented by which kind of feature. The notions of scale and resolution govern the choice. For example, an application using small scale maps may represent towns as point features (represented by TPointF class), while on a larger scale they may be represented by area features (represented by TSurfaceF class). To comply with this presumption, the definition of ADTFeature class must consider the specialized classes of ADTGeometry and ADTTheme. Molenaar (1993) discusses this issue in detail.

The ADTFeature object created at this stage has to be considered incomplete, because of the lack of geometric content (see Figure 6.9). Completion can only occur when the reference to the geometric container has been established and the geometric container filled with all necessary references to ADTGeometry objects (see Figure 6.10).

The ADTGeometry objects are the last to be created. The reason for this is that ADTFeature objects are not georeferenced before the stage of data acquisition. The specialized class of ADTFeature defines the specialized class of ADTGeometry object to be captured. When the user decides that a river will be represented by a line feature, the ADTGeometry objects to be captured are of the TArc class, and certainly not of the TTriangle class. Because the TArc object has references to two TNode objects, the user is forced to capture (create) two TNode objects prior to the creation of the TArc object. This engenders strict discipline in collecting data with expected high consistency.

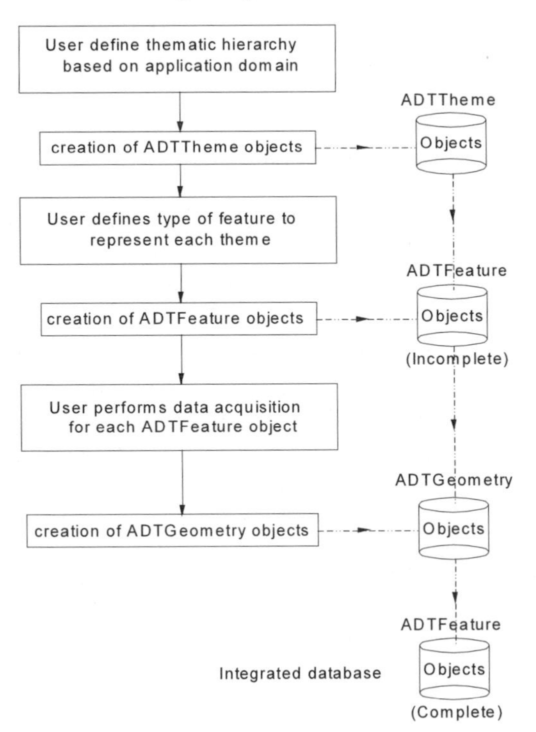

Fig. 6.9 Steps to creating objects.

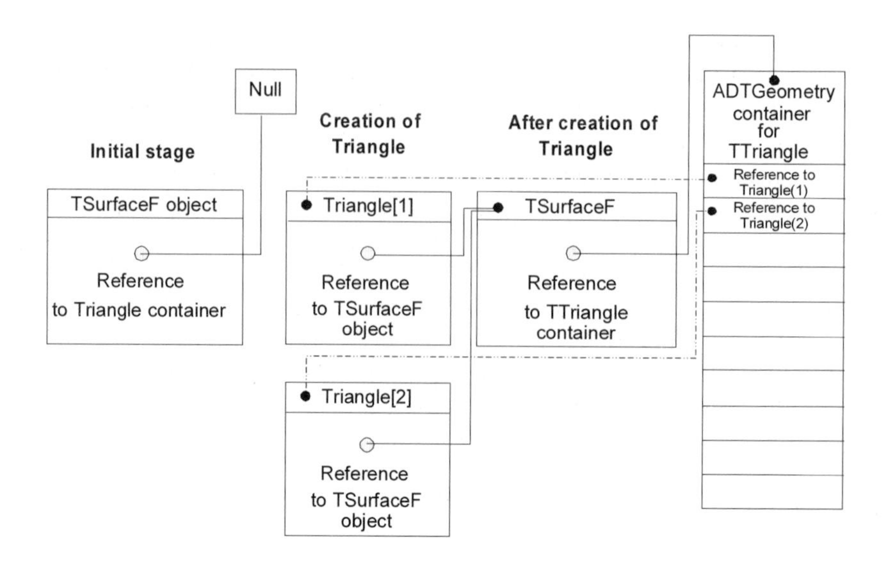

Fig. 6.10 Referencing scheme

6.2.6 Behaviour of Objects in the Database

By defining the hierarchies and relationships between the objects as outlined above, every object can respond to the message it receives from another object (whether self-activated or not). This kind of operation is efficient and consistent, since the appropriate operation is specific. For example, a user may wish to display the area features in a perspective view. Using a broadcasting mechanism, the user sends a message, such as 'Draw-3D,' to all objects belonging to the class TSurfaceF. On receiving this message, each TSurfaceF object then reacts to it by sending another 'Draw-3D' message to all of its component TTriangle objects (by searching in its geometric container). After each TTriangle object has received the message, it is sent to all three vertices, the TNode objects. The message asks the TNode objects to make a perspective transformation and then, using the transformed coordinates, to draw straight lines between themselves, perhaps adding colour-fill or shading if so requested.

Considering the aspect of spatial access, we observe that TLineF and TSurfaceF objects include references to geometric containers of classes TArc and TTriangle respectively. Taking a TSurfaceF as an example, and given a spatial location, the spatial search operation can be coarse to fine using, for example, the bounding rectangle of the TSurfaceF object as spatial index. The containment test is then performed in a simplified and fast manner. On receiving a positive result, the spatial search is then limited to TTriangle objects which are components of the TSurfaceF object. The search can then be performed using a reference to the TTriangle container that is one component of the TSurfaceF. This TTriangle container, which is specific to the TSurfaceF, contains a series of references to TTriangle objects. The references to the object offer a fast way of accessing the object component, that is, the attributes and methods.

Regarding the interfacing of system and user, the objects of classes under the ADTFeature hierarchy should provide all the necessary interfaces. During the database operation (deleting, modifying and so on), the objects of classes under the ADTGeometry hierarchy should not be directly accessible to users and should be under the complete control of each specialized ADTFeature object.

The example of implementation given as C++ object-oriented programming language is presented in Pilouk (1996) and Abdul-Rahman (2000). The focus is on the aspects of object creation, dynamic referencing and inheriting. For simplicity, a fixed size array is chosen as the container of references to each specific ADTGeometry object. Other versions of C++ offer more powerful container class libraries which can be used for the real implementation. We have implemented part of this definition in our software ISNAP (Integrated Simplicial Network Applications Package). The experimental investigation has demonstrated the feasibility of the design, thereby stimulating further investigation into the matter, for example response time and efficiency in spatial search operation. The implementation of this logical design in a commercial OODBMS environment still needs further exploration.

6.2.7 Comparison with Other OO Approaches

In comparison with the approaches presented by Webster and Omare (1991), Worboys et al (1990), Kainz and Shahriari (1993), the object-oriented approach discussed here offers a more flexible structure where users have the freedom to select different types of geometric representation per thematic class with respect to the scale of data acquisition. As an example, a city can be represented by an area object when the data is acquired from a map of scale 1: 50000, or a point object if acquired from a map of 1: 500000. The other approaches mentioned above have only adopted the inheritance hierarchies. For example, Webster and Omare (1991) defined a point feature as a supertype (parent or ascendant) of the node class, where a geometric class is a subtype (child or descendant) of a thematic class, which is similar to the approach used by Kainz and Shahriari (1993). Worboys et al., (1990) defined a district class as a specialization (child) of a polygon class where a feature object class is a child of a geometric class. In both cases, the consequence is that only one type of geometric representation is allowed in a hierarchy of that feature object. This restriction might prove too stringent and so the rigid inheritance approach can only be used as a logical design for a particular application. The whole object hierarchy has to be redefined when the database has to be upgraded to use a more precise geometry.

The rigid inheritance approach described above may not be suitable for UNS. In UNS, the possibility of having multiple geometric representations per class helps minimize the number of features stored in the database. The approach suggested in section 5.2 of using the aggregated hierarchy permits the selection of the type of representation according to whatever is available at the time of data acquisition. For instance, there may be many cities on the map with different possible representations, such as point and area, depending on their sizes. If the application only needs cities to be represented as points, each point object can be derived from the area object residing in the database, for example by using a cartographic generalization function of that area object.

6.3 Discussion

This chapter has presented thus far the translation from the IDM into a relational and an object-oriented UNS. Note, however, that only the necessary attributes are included in the relational UNS. This approach offers a quick and simple way of implementing an integrated database. Although good performance in terms of response time may not be obtained, the realization of all the necessary relationships between the data elements is facilitated. The control of database consistency depends on this minimum set of relationships between the data elements, even when the object-oriented approach is used. Significant performance gain in terms of response time is expected from the object-oriented approach, because links and pointers are used for navigating in the database instead of Cartesian products, as in the relational approach. Joining several tables together results in a long response time during a data retrieval and search operation in a large database. Most relational DBMSs offer a simple solution by creating an index file that can be thought of as a reordering of the records, using criteria on a selected column of each table. A typical indexing method is a binary tree (B-tree, Knuth, 1973), which may not, however, be suitable for indexing spatial data. The object-oriented approach permits the implementation of a more suitable spatial indexing method, such as Grid File, R-tree (Guttman, 1984; Oosterom, 1990). This method, however, requires a greater effort in implementing the index structure. Some DBMSs, such as Illustra (1994), claim to have offered a solution by providing R-tree to support efficient access to spatial data. Note that the index structure provides additional relationships among data elements. Most of these relationships can be inferred from the minimum set of relationships obtained from normalization in the relational approach and so they may be considered redundant, which asks for special care during the updating of the database.

It is worth mentioning that a UNS derived from the IDM can be managed by a single DBMS. Users only deal with one system and one user-interface. The time required for studying the use of the different commands of different databases (even for the same kind of operations) can be reduced, allowing users to concentrate on the actual application.

Chapter 7 OBJECT-ORIENTATION OF TINs SPATIAL DATA

7.1 Introduction

The capabilities of object-oriented (OO) techniques have in recent years presented a very promising tool for the development of information systems, especially those requiring the implementation of complex data modelling. OO programming techniques are now being applied widely. OO programming has tremendous potential; GIS is one example. OO techniques of programming and design promise to produce easier to maintain software for less effort and expense (Ross *et al*, 1992). Conventional software development suffers from a number of drawbacks such as endless lines of code, while OO programming allows programmers to build an application program by using existing or easy-to-build entities called objects (object - the term used in OO programming for, an instance of a constructed class). Therefore, it seems natural to apply OO techniques for geo-scientific computations such as TIN spatial data modelling.

This chapter provides descriptions of TIN tessellations and spatial data modelling using OO techniques. Section 7.2 discusses the concepts of OO. OO design for TIN tessellations is discussed in section 7.3. A discussion on the development of TIN spatial data modelling is provided in section 7.4, and the POET OO DBMS development in section 7.5. The development of OO TIN-based systems for GIS follows in section 7.6 and the chapter closes with a summary. The implementation of OO techniques for TIN data tessellations has been further discussed in Abdul-Rahman (1999). Further implementation using an OO database management system (DBMS) is described in Abdul-Rahman and Drummond (2000).

7.2 Object-oriented Concepts

Object-oriented conceptual modelling is now widely utilised in many fields including GIS. The concepts of OO such as object classification, encapsulation, inheritance, and polymorphism have made modelling of complex real world objects easier. As mentioned above the object-oriented approach is now being promoted as the most appropriate method for modelling complex situations that are concerned with real-world phenomena, and thus applicable to GIS. Object-oriented concepts are considered more flexible and powerful than the traditional structural programming and other major database models such as the relational or entity-relationship model. Object-oriented concepts contribute to modelling as follows:

- Considering objects and abstraction mechanisms (classification, generation, aggregation and association), these aspects of OO can be used for modelling real world phenomena, e.g. modelling of spatial data for geoinformation systems; and
- Considering inheritance, propagation, encapsulation, persistence, Abstract Data Type (ADT), polymorphism and overloading, these aspects of OO can be used to construct and implement the model discussed in (a).

The usefulness of these concepts in spatial modelling is explained in the following section.

7.2.1 The Abstraction Mechanisms

Data abstraction is a method of modelling data. Object-oriented design uses four major abstraction mechanisms: (1) classification, (2) generalization, (3) inheritance, and (4) polymorphism. In object-oriented programming, any physical or logical entity in the model is an "object". The definition of a type of object is called a "class", and each particular object of that type known as an "instance" of the class. Once a class has been defined, it can, potentially be reused in other programs by simply including the class definition in the new program. However, it is not necessary for the programmer who uses a class to know how it works, s/hey simply needs to know how to use it. The definition of operations on or between objects are called "methods", and the invocation of methods is referred to as "passing a message". Recent research in software engineering has promoted an object-oriented design method by which real world objects and their relevant operations are modelled in a program which is more flexible and better suited to describe complex real world situations (Khoshafian and Abnous, 1995). Object orientation also may be considered as a particular view of the world which attempts to model reality as closely as possible (Webster, 1990). Details of all relevant OO concepts (object, abstraction, data types, class hierarchy, inheritance, classification, aggregation, generalization and association) can be found in the OO literature such as Booch (1990), Bhalla (1991), and Stroustrup (1997). The following are some OO terms:

Classification

Classification can be expressed as the mapping of several objects (instances) onto a common class. In the object-oriented approach, every object is an instance of a class (a class is a fundamental building block in an OO language). It describes common features of a set of objects with the same characteristics; a class also defines nature of a state and behaviour, while an object records the identity and state of one particular instance of a class. Abstract Data Type (ADT) is the name of the mechanism to create a class of spatial objects or any class in a domain of objects. An object is a basic run-time entity in an object-oriented system. This entity includes data and procedures that operate on data. Viewed from a programming stand point, objects are the elements of an OO programming system sending and receiving messages.

Generalization

Generalization in OO provides for the grouping of classes of objects, which have some operations in common, into a more general superclass. Objects of superclass and subclass are related by an *"is a"*- relation, since the object of a subclass also "is-a" (n) instance of a superclass.

Inheritance

Inheritance allows the building of a hierarchy of types or classes that best describes the real world situation in the application field. Each class can take all or part of the structural or behavioural features from other classes, which are its parents. In turn, the newly defined class is a child of the classes from which it has inherited its features. Inheritance helps in deriving application-oriented classes without starting every definition from scratch. Also, it makes it easier to create logically complex classes from simpler classes.

Polymorphism

Polymorphism is a mechanism to define the different actions of the same named function on different classes. It is implemented by inheriting some functions from parent classes and overriding or modifying part of them. Usually, the newly created class has similar but not the same behaviour as its parents for that functional aspect. Polymorphism provides great flexibility in class derivation, for example, the *calculate_perimeter* operation may have different implementations for different classes such as class "area", class "triangle", class "polygon", etc. Each class performs the calculate perimeter operation differently although it has the same function name.

7.2.2 The Programming Language

Object-oriented concepts were originally developed in early programming languages such as Simula in 1960's. Other OO programming languages such as Smalltalk, C++ and Java have also been developed since then. Although Java is said to be widely used for the Internet or distributed computing environment these days, C++ language is much more widely used and offers more OO concepts than other languages (Stroustrup, 1997). There are several C++ compilers available from major software/compiler vendors for a wide variety of computer systems. Most of these compilers are meant for a wide variety of scientific computing tasks, including for instance geoinformation modelling and computations. In the work reported in this book, the Borland™ C++ compiler was utilised for all the software development.

7.3 Object-oriented TIN Tessellations

OO TIN tessellations software has been developed for the construction of 2D and 3D TIN data structures. The algorithms are described in Chapter 8. The descriptions of the OO TIN tessellations are presented in the following segment.

7.3.1 Classes for 2D TIN Tessellations

Using the above OO mechanisms, the spatial tessellations are designed as shown in Figure 7.1. In this design, the Booch (1990) notation was used to represent the hierarchy of the classes. Booch has provided one of the techniques for designing class hierarchy. Other possible techniques are notations such as those of Rumbaugh and the Unified Modelling Language (UML). In two-dimensional (2D) spatial tessellation, four major classes have been recognised, the classes are TDistanceTransformation (TDT) class, TVoronoiTessellation (TVor) class, TTinGeneration (TTinGen) class, and TTinView (TTinView) class (see Figure 7.1). The TDT class is used to calculate and generate a distance transformed image of given object pixels. The TVor class is used to generate the Voronoi image of the object pixels. The corresponding TIN of the object pixels can be determined by using the TTinGen class, and the TIN viewing is handled by the TTin-View class.

In this work, not all OO mechanisms were used. The two most useful mechanisms are classification, and inheritance. The following section describing all the relevant classes the spatial tessellations.

The Class *TDistanceTransformation* generates a distance transformed image from a given rasterised data set. Operations or methods in this class are: *SetBackground, GetUpperMask*, *GetLowerMask*, *ForwardPass*, and *BackwardPass.* The details of these methods or procedures were fully described in Abdul-Rahman and Drummond (1998, 1999). Here, only their relationships with other classes in the class hierarchy are described. The details (class headers which includes all the related attributes and methods) for each class are presented in Abdul-Rahman (2000).

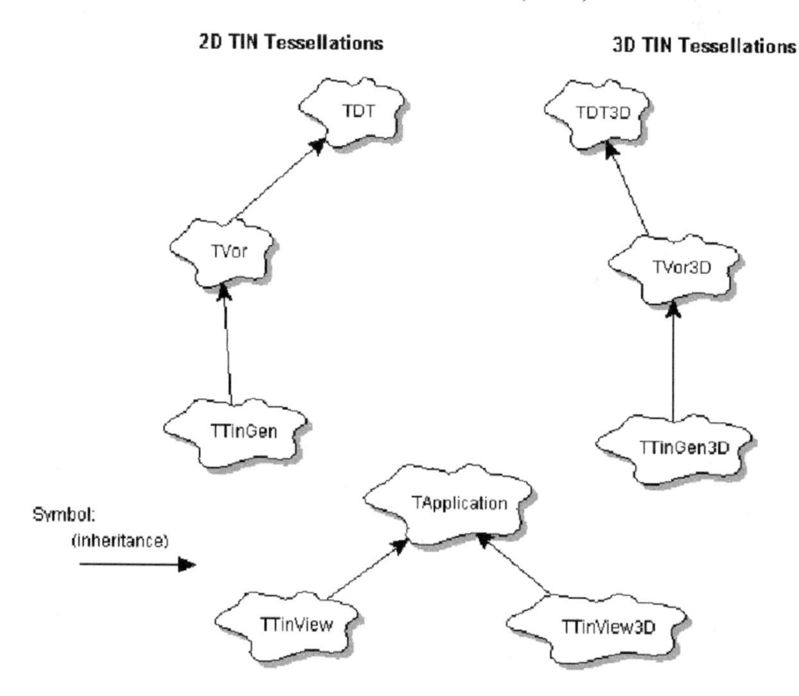

Fig. 7.1 The classes hierarchy for the 2D and 3D TIN tessellations

The following class *TVoronoiTessellation* generates a Voronoi image from a given distance transformed data set. The major methods in this class are *ForwardVoronoi* and *BackwardVoronoi.* These two operations are to generate the tessellated image in two passes. The forward pass begins from the top left corner of the image while the backward pass works reversely (i.e. from the bottom-right pixel to the top-left pixel). The class mentioned above, *TTinGeneration* produces a TIN from a given Voronoi image data set. The *ScanlinesUp* and *ScanlinesDown* methods are to detect the TIN's triangles from the Voronoi images. After having generated the TIN then, the next task is to display (visualize) them. The visualization make uses of the Borland's C++ compiler predefined class *TApplication*, that is the superclass for the TTinView.

- The following gives the definitions of the 2D TIN classes:

```
class DistanceTransform
   {
   public:
   // member data
   typedef struct MpiStruct
   {
   short Nscanlines;
   short Npixels;
   ...
   } MpiType;

   // member functions
   DistanceTransform();          // constructor

   void SetBackground(ImagePPtr Pixel, int Bg, int Fg);
   void GetUpperMask(int r, int c, ImagePPtr Pixel, Mask& MaskPix);
   void GetLowerMask(int r, int c, ImagePPtr Pixel, Mask& MaskPix);
   void DistancePassOne(ImagePPtr Pixel);
   void DistancePassTwo(ImagePPtr Pixel);
   void ForwardDistance();
   void BackwardDistance();
   ~DistanceTransform();         // destructor
   };

class TVoronoiTessellation : public TDistanceTransform
   {
   public:
   ...
    // Function members
   void CopyImage(ImagePPtr, ImagePPtr&);
   void SetBackground(ImagePPtr, int, int);
```

```
  void GetUpperMaskDist(int, int, ImagePPtr, Mask&);
  void GetLowerMaskDist(int, int, ImagePPtr, Mask&);
  void GetUpperMaskVoronoi(int, int, ImagePPtr, Mask&);
  void GetLowerMaskVoronoi(int, int, ImagePPtr, Mask&);
  void ForwardPass(ImagePPtr, ImagePPtr);
  void BackwardPass(ImagePPtr, ImagePPtr);
  void ForwardVoronoi();
  void BackwardVoronoi();
  };

class TINGeneration
  {
  public:
  typedef struct MpiStruct
  {
  short Nscanlines;     // no. of image rows
  short Npixels;        // no. of image columns
   ...
  } MpiType;

  typedef struct VertexStruct
  {
  DataType N1;
   ...
  } TVertex;

  typedef struct TsNodeStruct
  {
  short x;
   ...
  } TsNode;

  // function members
  void GetMask(int, int, ImagePPtr, Mask&);
  void GetSubImage(int, int, ImagePPtr, Mask&);
  void ScanlinesUp(Mask);
  void ScanlinesDown(Mask);
  void Scanlines(ImagePPtr);
  void MakeTIN();
   ...
  };
```

7.3.2 Classes for 3D TIN Tessellations

The tessellations of the 3D TIN have also been developed; their class hierarchies are very similar to that of 2D TIN version. The only difference is the computation dimension (the additional third dimension), and the way to visualize the generated 3D tessellation files. The 3D tessellations also have four major classes. The classes are *TDistanceTransformation3D* (TDT3D), *TVoronoi3D* (TVor3D), *TTinGeneration3D* (TTinGen3D), and *TTinView3D* (TTinV3D) (refers to Figure 7.1). Detailed definitions of each class are presented in Abdul-Rahman (2000).

For purpose of displaying the DT and the Voronoi images, the ILWIS™(1996) and AVS™(1997) packages have been utilised. The later package is for 3D images while the former is for 2D images.

7.4 Object-oriented TINs Spatial Data Modelling

In this section, we provide a discussion of the OO TIN spatial data modelling techniques. The general modelling steps (recall Figure 4.2) can be considered to describe TIN spatial data modelling. That is, the three-step approach, namely the conceptual, the logical, and the physical steps. In this work, the conceptual step is implemented by utilising TIN as a method to represent spatial objects. Spatial objects are perceived as TINs. Then, by having the TINs data constructed, a model to describe the objects (i.e. their connectedness between objects) can be established. The description of the objects and how they relate to each other, for example from TIN nodes to TIN surface is the logical step. All the created classes (Node, Edge, Area, and Body) are OO techniques are then physically modelled in OO database environment.

The class schema for spatial data modelling is described below.

7.4.1 The Class Schema

The schema is based on several classes, they are Spatial Objects (the super class), and four major subclasses which are Node, Edge, Polygon, and Solid.

Spatial objects

The spatial object class is a general class of the real world objects. It is the super class in the class hierarchy. In this work, an assumption is made that all other objects are derived from the superclass (TSpatial object) (see Figure 7.2). All terrain objects could be categorised into several sub classes such as points, lines, areas, and solids (volume) features. In OO modelling, these feature types are the classes in the modelling hierarchy.

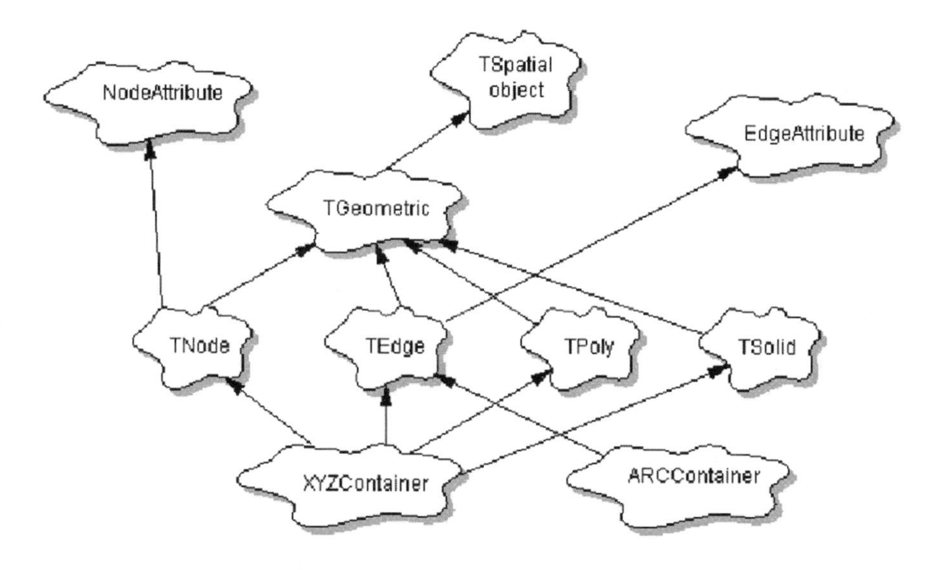

Fig. 7.2 The class diagram (using the Booch notation)

Node

A node can be considered as the most basic geometrical unit in spatial data modelling. It may represent point entities or point objects at a particular mapping scale. Examples of point objects are wells, terrain spot heights, and the like. In geoinformation, we may represent these objects by a class called a node class. The coordinates of the nodes (including the nodes representing edges) are held by a coordinates container class, called XYZContainer class.

Edge

An edge can be represented by two nodes at each end (i.e., a start node and an end node). In this study, we consider two end points make a straight edge. This edge type can be used to represent linear features. The arc container class, called ARCContainer holds all the arcs. The arc containers also serve any other class which requires arcs data in their operations, for example the polygon class needs the arcs in order to form polygons.

Polygon

A polygon (sometimes known as a surface) is used to represent area features such as lakes, ponds, etc. A polygon may be constructed by chains of closed edges.

Solid (or Body)

A solid is a representation for solid or body features such as buildings, or trees. A chain of points and lines form body objects. A 3D TIN can be represented by a series of triangle nodes and edges as mentioned in the previous chapter.

The class schema in Figure 7.2, depicted using Booch (1990) notation is the representation of the TIN spatial data model. The schema has four geometric classes, namely *TNode, TEdge*, *TPoly*, and *TSolid* and two types of containers: geometry and attribute. The geometric containers contain the XYZ locations (held by the *XYZContainer*) and the *ARCContainer* whereas the attribute containers are for the thematic values, e.g. names. The attribute information is held by the *NodeAttrbute,* and the *EdgeAttribute.*

The following gives the definitions of the classes as presented in Figure 7.2. In Booch notation, each class is represented by a "cloud-look" diagram. It contains data and methods for a particular class. The arrow shows the link between a class with another classes. More detailed class definitions are given in Abdul-Rahman (2000).

The geometric classes are:

```
class TNode
    {
    public:
        struct XYZContainer
        {
        double x;
        double y;
        double z;
        };
    XYZContainer Point[maxpoint];

    struct NodeAtrContainer
    {
    int NodeNum;
```

```
    string NodeName;
    };
    NodeAtrContainer NodeAtr[maxnodename];
    ...
    ...

    void GetXYZCoordinates();
    void Get2Node();
    void NodeAttribute();
    };

class TEdge
    {
    public:
            ARCContainer Arc[maxarc];
            EdgeNameContainer-  EdgeAtr[maxarcname];
            ...
            ...
            void ReadARCs();
            void GetArcLength();
            void GetArcAttribute();
            int CheckQuadrant(float, float);
            float Bearing(float, float, float, float);
            float GetArcAzimuth(float, float, float, float);
            void EdgeAttribute();
    };

class TPoly
    {
    public:
    struct TINContainer
    {
            int Node1;
            int Node2;
            int Node3;
    };
    TINContainer Triangle[maxtriangle];

    struct TENContainer
            {
            int TriNum;
            int NumofNbr;
```

```
                int Nbr1;
                int Nbr2;
                int Nbr3;
                };
        TENContainer TINNbr[maxtriangle];

        void ReadTINs();
        void GetTINNeighbour();
        void GetTINNodes(int,
                double&, double&, double&,
                double&, double&, double&,
                double&, double&, double&);
        float GetTINArea();
        void GetPolyArea(int, int, double&);
        };

    class TSolid
        {
        public:
        struct TENContainer
                {
                int Node1;
                int Node2;
                int Node3;
                int Node4;
                };
        TENContainer TEN[maxtriangle];

        void ReadTENs();
        void Get3TINNodes(int,
                        double&, double&, double&,
                        double&, double&, double&,
                        double&, double&, double&,
                        double&, double&, double&);

        float GetVolume(double, double, double,
                        double, double, double,
                        double, double, double,
                        double, double, double);
        };

class XYZContainer
```

```
        {
        public:
        double x;
        double y;
        double z;

    XYZContainer() {}
    ~XYZContainer() {}
        };

class TINContainer
        {
        public:
    int Node1;
    int Node2;
    int Node3;
        };

class TENPolyContainer
        {
        public:
    int TriNum;
    int NumofNbr;
    int Nbr1;
    int Nbr2;
    int Nbr3;
        };

class TENContainer
        {
        public:
    int Node1;
    int Node2;
    int Node3;
    int Node4;
        };


    class ARCContainer
        {
        public:
        int StartNode;
```

```
int EndNode;

ARCContainer();
~ARCContainer();
};
```

- The attribute classes are:

```
class EdgeNameContainer
    {
    public:
    int EdgeNum;
    char EdgeName[30];
    };

class NodeNameContainer
    {
    public:
int NodeNum;
 char NodeName[30];
    };
```

7.5 Object-oriented TIN Spatial Database Development

This section explains the development of an OO database for TIN data using a commercial database management system, POET OO DBMS.

7.5.1 The POET OO DBMS

The POET (Persistence Object and Extended Technology) DBMS is utilised in this work as part of the OO spatial data modelling. The package works under Windows 95 operating system for the PC environment, that is the major computing environment adopted in this work. The package is also chosen due to its economic reason; it costs less and can work with Borland C++ programming language, the language adopted for the entire software development in this book. The database package is said to have the following capabilities (POET, 1996):

a. Encapsulation,

b. Inheritance, and
c. User-defined data types.

These properties are among the important OO features that can be useful for our TIN spatial data modelling purposes.

7.5.2 The POET Database Schema

The DBMS is used to generate OO database from the constructed TIN spatial data. In this work, the schema needs to be modelled according to the POET database model (POET, 1996); it is required that all the C++ classes are constructed as classes which POET can understand. In this case, all the classes in the schema have to be compiled by POET PTXX compiler as shown in Figure 7.3. The PTXX compiler maps all the normal C++ classes into several relevant PTXX schema files which in turn are used for writing application programs (running under the normal C++ compiler) as well as for populating the database. The PTXX compiler also generates the OO database from the schema.

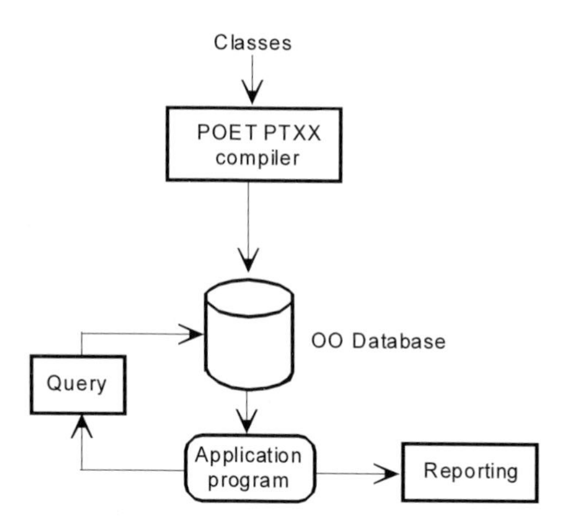

Fig. 7.3 The POET database development flow.

7.5.3 The POET Database Browser

In POET, once the database schema has been properly compiled, then the generated database can be inspected by using the built-in browser. All generated objects can be examined for further database operations. Figure 7.4 illustrates the screen shot of the POET Developer module where the TIN database is developed.

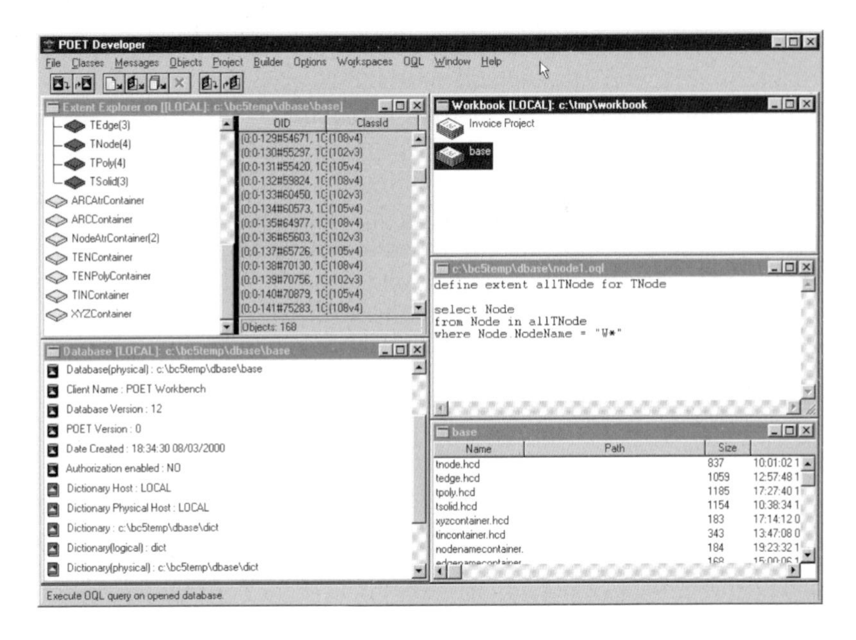

Fig. 7.4 The POET Developer which was used to develop TIN OO database and support database retrieval (query)

7.5.4 POET Database Query

The generated database can be queried by using a built-in database query facility within the POET Developer module. This built-in technique is adopted for this work. A query language similar to the Structured Query Language (SQL) can be utilised. Here the language is called Object Query

Language (OQL) and detailed syntax of the language can be found in POET(1996). The following is an example of a query which can be performed from the database:

```
defined extent allTEdge for TEdge;
select Edge
from Edge in allTEdge
where Edge.EdgeAtr.EdgeName = “River*”
```

In order to be able to manipulate the database, an application program has been developed. This program runs under the normal Borland C++ compiler but it makes use of the files which are generated by the POET PTXX compiler.

7.6 Object-oriented TIN-based Subsystems for GIS

The OO TIN GIS is based on several fundamental concepts and aspects of spatial data which have been discussed in the previous sections. The basic components of the system are data input processing, TIN data construction, TIN database, transformation operations, data output and user-interface. Rasterization forms a major operation in the data input component. Figure 7.5 shows the other major component of the proposed system - visualization, which uses the commercial software, i.e. ILWIS™ (Integrated Land and Water Information System) and AVS™ (Advanced Visualization System). These two packages are only used for display purposes, more especially for validating the output from the rasterization process. A simple user-interface as part of the software development is also developed. Besides the programs written for databasing purposes, a commercial database package is also used, called POET™ OODBMS as mentioned eralier. The DBMS package is for the development of the OO TIN spatial database.

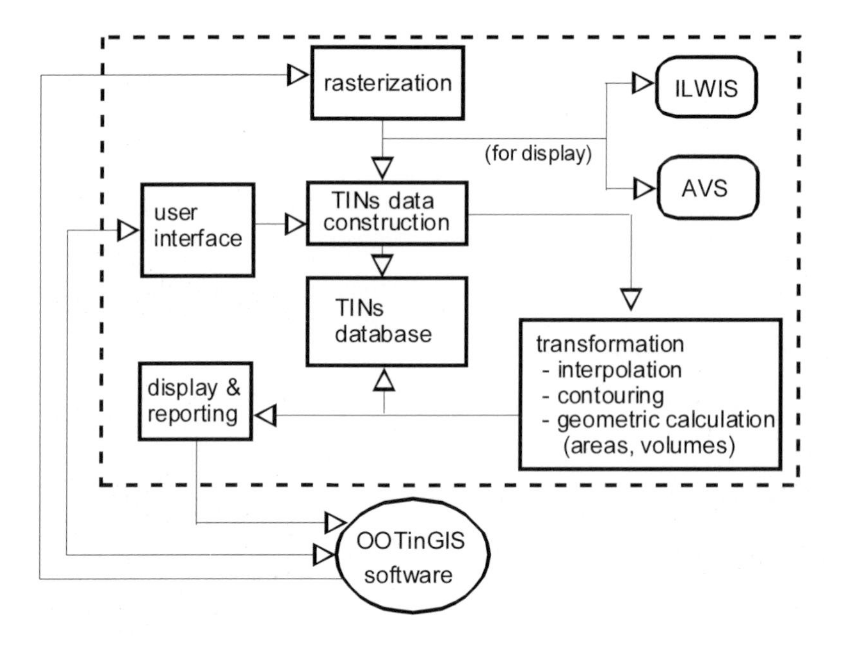

Fig. 7.5 The proposed system for the TIN-based spatial data.

7.7 Summary

This chapter has introduced the implementation of object-oriented techniques for TIN (2D and 3D) spatial data. The chapter reveals the usefulness of a commercial OO DBMS package to develop TIN spatial database schema as described in section 7.5. The development of software which has been described in this chapter could also be applied to a much larger system.

2D and 3D TIN tessellations are one of the major components of the work described in this book. These tessellations are shown to work perfectly in the OO environment. The approach described in this chapter can be implemented in the development of TIN-based GIS system. The graphic output of the tessellations shown in Figure 7.6 clearly demonstrates the workability of the OO technique. More results of the subsystems are presented in later chapter.

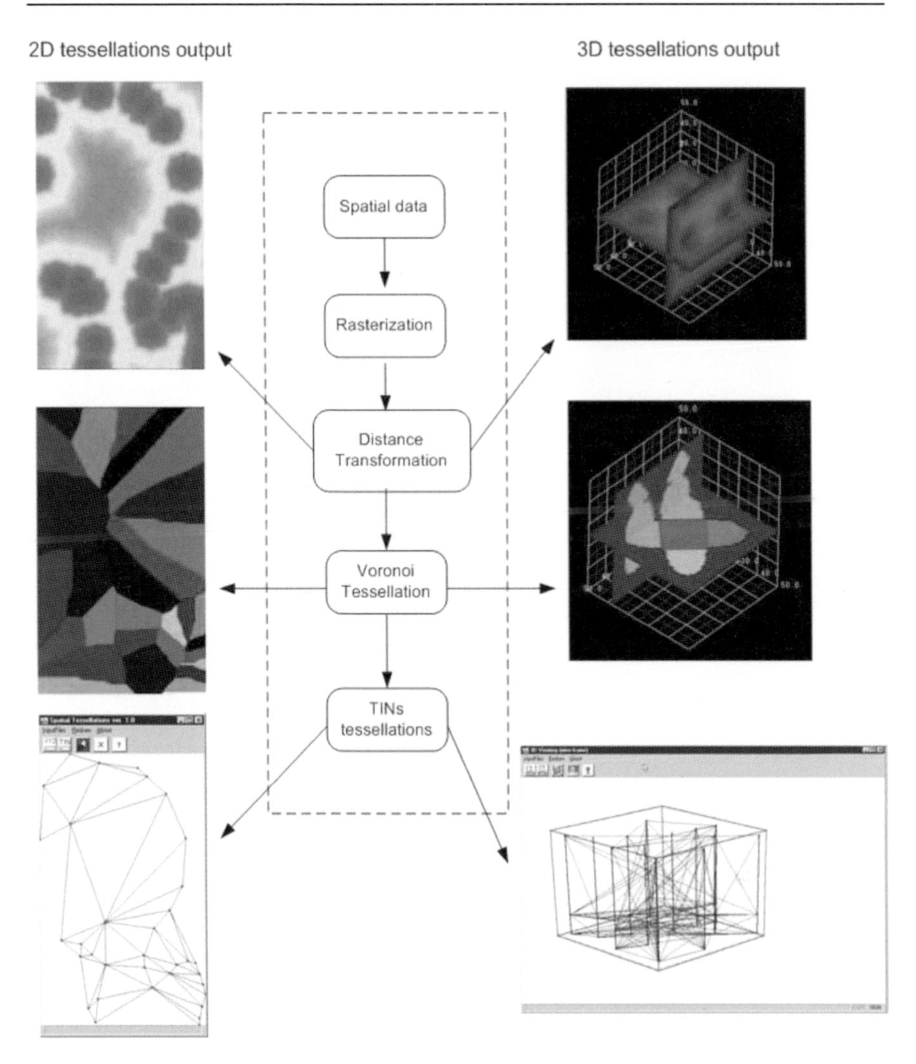

Fig. 7.6 The 2D and 3D TIN tessellations

Chapter 8 THE SUPPORTING ALGORITHMS

8.1 Introduction

This chapter introduces several major algorithms for TIN spatial data structuring and constructions. Data structuring for terrain surfaces has been investigated for several decades. The main concern of the earlier investigations was the suitability and the adaptability of data structures for representing terrain surfaces. A triangular irregular network (TIN) data structure was first presented by Peucker *et al.*, 1978. Several methods and techniques have since emerged for the construction of TIN structures (McCullagh and Ross, 1980; Watson, 1981; Mirante and Weingarten, 1982). Most of the techniques were attributed to Delaunay (1934) and known as Delaunay triangulation. TINs could be constructed either in the vector or in the raster domain. In this research, a raster technique for the construction of the TINs (2D and 3D) is used.

In this chapter, the algorithms for the construction of 2D TIN and 3D TIN spatial data will be introduced. These algorithms are named Distance Transformation (DT), Voronoi Tessellations, Triangulations, and Triangulations Data Structuring. In this work, visualization and rendering routines for 2D and 3D data have also been developed, as have rasterization programs for TIN data construction purposes. Each algorithm is explained in detail together with its C++ pseudo-code.

8.2 Distance Transformation

Originally, the term distance transformation (DT) was used by Rosenfeld and Pfaltz (1966) and later by Borgefors (1986). The DT was used to describe an operation of converting binary images to a grey-level image where all pixels have a value corresponding to the distance to the nearest feature (or object) pixel. The same principle has also been applied in other areas of interest such as raster-based GIS and remote sensing (Gorte and Koolhoven, 1990). The DT provides a method for calculating the distance from every non-object element in a two-valued raster data set to the nearest object element of a set of object elements. The Borgefors DT technique is a fundamental step in this raster-based TIN development. The transformed image can be used to generate a Voronoi tessellated image, and then a set of triangles can be generated from that Voronoi tessellated image. Triangles generated from Voronoi polygons are sometimes known as the dual product of the Voronoi polygons (Gold, 1991; Fortune, 1992). Borgefors (1986) identified several types of DTs known as City block, Chessboard, Octagonal, Chamfer 3-4, Chamfer 5-7-11, and Euclidean. Each DT produces different output images and requires a different computation time. Borgefors suggested that Chamfer 3-4 can be used for generating distance transformed images due to its processing simplicity. Description of other DTs can be found in Borgefors (1986). It is not the intention of this section to compare all the DTs but rather to explain them and then use the most appropriate one (i.e. that is relatively easy to implement); a detailed explanation of the DT which used in the TIN development is described later in this section.

Distance Transformation (DT) is a technique used in the image processing community for a range of applications, one example is zone mapping (Borgefors, 1986). A zone of accumulated distances could be mapped from a rasterised point. This DT concept is used in this research and the technique is one of the fundamental steps in the construction of the triangulation. The task is to generate a distance-transformed image of object pixels. In a raster image, object pixels could be in the form of random points, digitized points, digitized lines, etc.

Figure 8.1 shows an example of several points whereas the DT image of the points are illustrated in Figure 8.2.

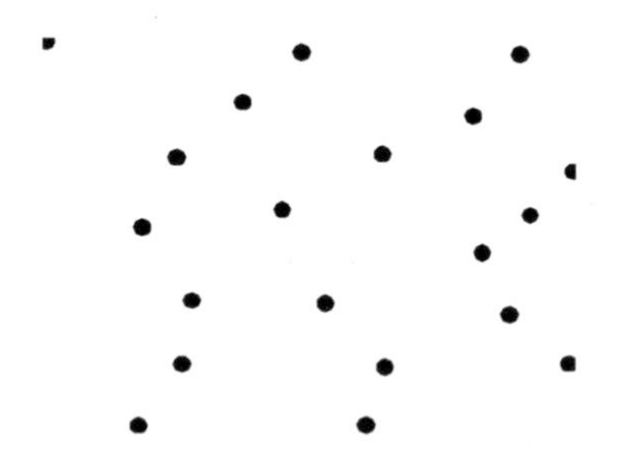

Fig. 8.1 Several kernel points (or object points)

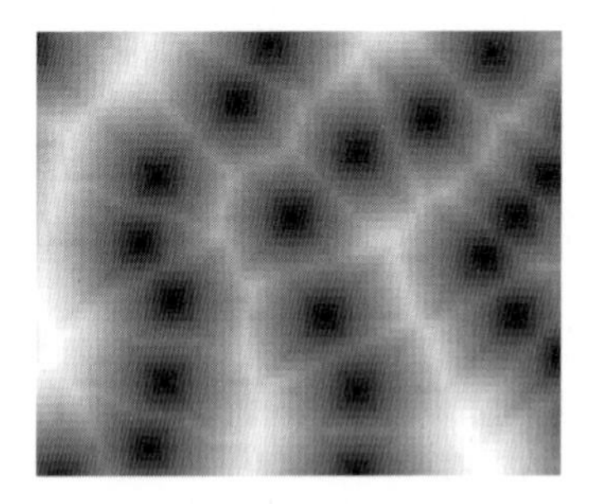

Fig. 8.2 The DT image of the several points as shown in Fig. 8.1

In Figure 8.2, the darkest spots represent the location of the kernel points. In the DT, each kernel point is used to generate distance image from neighbouring kernel points. Distances accumulate from the centre of the kernel points. In the above example, the centre of the kernel points gets the value zero (the darkest shade) and the distances gradually increase from the centre (indicated with brighter images) as shown in Figure 8.2. To perform the DT to an image of rasterised points, for example, a mask (or a window of 3 x 3 pixels) is required, as shown in Figure 8.3(a). The mask has 9 pixels (3 x 3 pixels). This mask is divided into two, called the upper mask, and the lower masks, as shown in Figure 8.3(d).

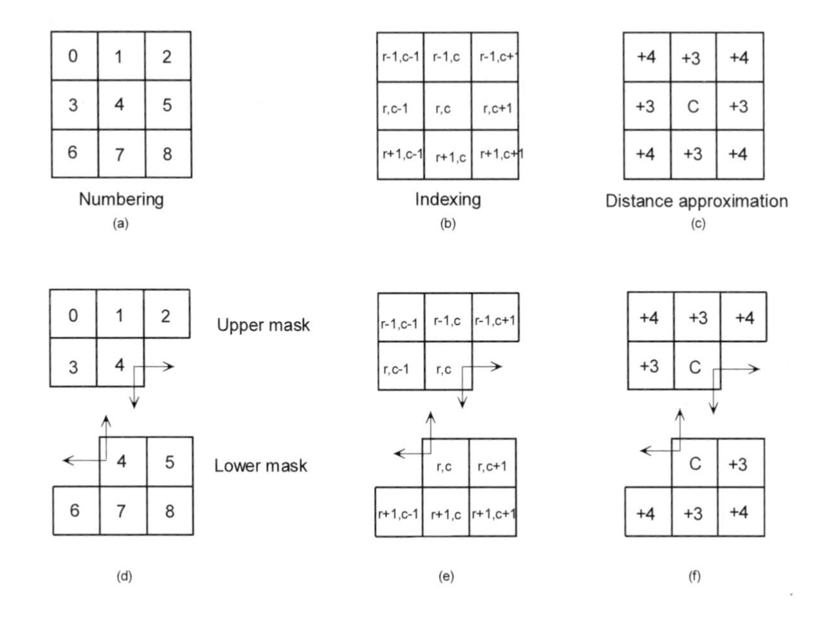

Fig. 8.3 Masks for the DT operations

The algorithm works with two passes of the entire image. The first pass (or scan) uses the upper mask while the lower mask is used for the second pass. Each pixel in the mask is indexed according to Figure 8.3(b) where the centre pixel of the mask represents the image pixel then being scanned.

In this algorithm, the DT works as follows: all object pixels are changed to zero (i.e. a value 0) and the rest of the pixels (i.e. the background pixels) to the highest possible value, e.g. an integer value of 32767 (of 16-bit data type architecture). The entire image is scanned in two passes using the Chamfer 3-4 mask of the Borgefors DT (Borgefors, 1986) as illustrated in Figure 8.3(c). The first pass (scans with upper-mask) begins from the first pixel (i.e. the top-left pixel) and goes to the last pixel of the image. In the first pass, all the pixels which are covered by the mask get a new value. Each pixel's value has added to it either a value of 3 or 4 depending on the location of the pixel. Then, the minimum value is determined from the five possible candidates and assigned to the current pixel location. The mask is then moved to the next pixel location. At this next location, the minimum value for this pixel is again determined and assigned.

This process continues to the last pixel location (i.e. the bottom-right pixel) of the image. The result of the first pass operation is used for the second pass which operates in reverse order (i.e. from the last pixel to the first pixel of the image). It is a recursive operation. Finally, a DT image is generated after these two passes are carried out. In this DT image, all pixels contain the approximate distance to the nearest kernel points (object pixels).

The following pseudo-code describes the DT algorithm:

```
// Procedure to Set the background image
void set background()
 {
  Set loop for rows (first row to last row)
   {
    Set loop for columns (first column to last column
     {
      if (Pixel value not equal to background)
         Set Pixel value to zero;
      else
         Set Pixel value to background (highest possible value);
     }
   }
}
// Procedure to assign the Upper Mask
void GetUpperMask()
 {
  Assign the Mask[0] to Mask[4] to the corresponding pixel lo-
cations,
         e.g., Mask Pixel[0] = Pixel at [row-1][column-1];
 }

// Procedure to assign the Lower Mask
void GetLowerMask()
 {
  Assign the Mask[4] to Mask[8] to the corresponding pixel lo-
cations,
         e.g., Mask Pixel[4] = Pixel at [row][column];
```

```
    }
  // Procedure to compute distance in forward pass
  void ForwardPass()
    {
     Set loop for row(first row to last row)
       {
        Set loop for (first col to last col)
          {
           GetUpperMask();
           If Mask has odd index add 4 to the Mask value;
           else
           Add 3 to the Mask value;
             }
           Get the minimum value of Mask[0] to Mask[4] and assign
           to this pixel;
          }
        }
    }

  // Procedure to compute distance in backward pass
  void BackwardPass()
    {
     Set loop for row(from last row to first row)
       {
        Set loop for col(last col to first col)
          {
           GetLowerMask();
           If Mask has odd index add 4 to the Mask value;
           else
           Add 3 to the Mask value;
             }
           Get the minimum value of Mask[0] to Mask[4] and assign
           to this pixel;
          }
        }
}
```

The above steps then combined as follows into one main DT routine:

```
  // Procedure to compute the distance using forward and back-
 ward
  void Forward&Backward()
```

```
{
Reads the input Image;
Set the Background;
Compute distance using the FirstPass;
Compute distance using the SecondPass;
Write and save the transformed image to file;
}
```

An image of a DT for a number of points within a data set (kernel points) is illustrated in Figure 8.2. The darkest spots in the image represent the kernel points, and it gradually brightens outward from the points. The DT algorithm appears to work well.

8.3 Voronoi Tessellations

Voronoi polygons are also known as Thiessen or Dirichlet polygons. They have been considered one of the fundamental structures in computational geometry and other fields such as GIS. Voronoi polygons are often used in GIS as a method for analysing points data, for example for finding nearest neighbours (Burrough and McDonnel, 1998). In Voronoi polygons, one centroid point represents one polygon. The extent of each polygon indicates the influence of the centroid point with respect with the neighbouring points.

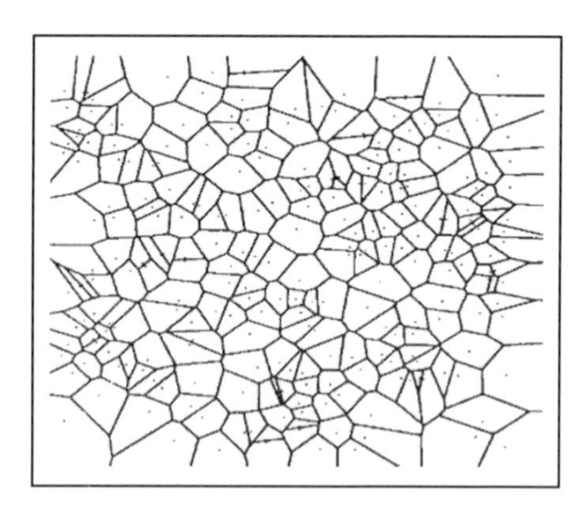

Fig. 8.4 Example of Voronoi polygons represented by several data points (after Fortune (1992).

This type of polygon is useful in GIS, e.g. for zone mapping or for determining the region of influence of a phenomenon or buffering (Gold *et al*,

1997). Figure 8.4 shows Voronoi polygons where each is represented by a centroid point.

Voronoi polygons can be constructed from DT image of kernel points as described in section 8.2. The generation of the polygons can be done either in parallel or in stages. In this algorithm, the tasks were carried out in parallel. If the DT generation as described in section 8.2 is re-examined, it involves three steps. First, change the object pixel value to zero (i.e. 0) and the background image to the highest possible value. Second, determine the minimum value of the current pixel location among five possible candidates of the upper mask. Third, assign the minimum value to the current pixel location. In other words, the pixel value represents a distance value of the pixel calculated from the nearby object pixels.

To generate the Voronoi-tessellated image parallel with DT operation, two output files are needed. That is one for the DT image and the other for the Voronoi image. Computing the DT image according to the algorithm describe in section 8.2 involves the following steps at a particular pixel [i, j]: First, the mask is "put" on the pre-processed image, the mask centre (having the value 0) at [i, j] of the pre-processed image. Secondly, the values of the mask are added to the values pixels that are being covered. Thirdly, the minimum of the 5 resulting values is determined and assigned to [i, j] of the current distance transform image. Before continuing to the next pixel, for which the distance is to be computed, the value for the second output image, the Voronoi tessellation image, at [i, j] has to be assigned. This is done by determining the location of the pixel where the minimum value was found just before, e.g. at [i, j-1]. The pixel value of the original image at [i, j] is then taken and assigned to [i, j] of the Voronoi tessellation image (see Figure 8.5 and also Figure 8.6). This method of computing the Voronoi and DT in parallel was also suggested by Borgefors (1986) that "the computing of the Voronoi tessellation image can be done by first computing distance transformation from an object pixel while at the same time keep track from which pixel the distance is computed".

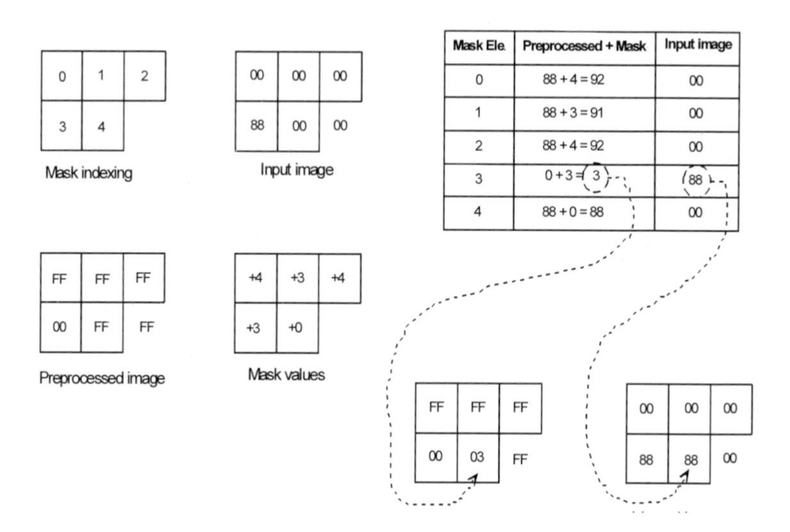

Fig. 8.5 DT computation and Voronoi image generation during the forward pass.

A more complete picture for the parallel process of the DT and Voronoi tessellation implementation is illustrated in Figure 8.6 where the outcome of the first pass and the second pass applied to the input pixels is clearly illustrated.

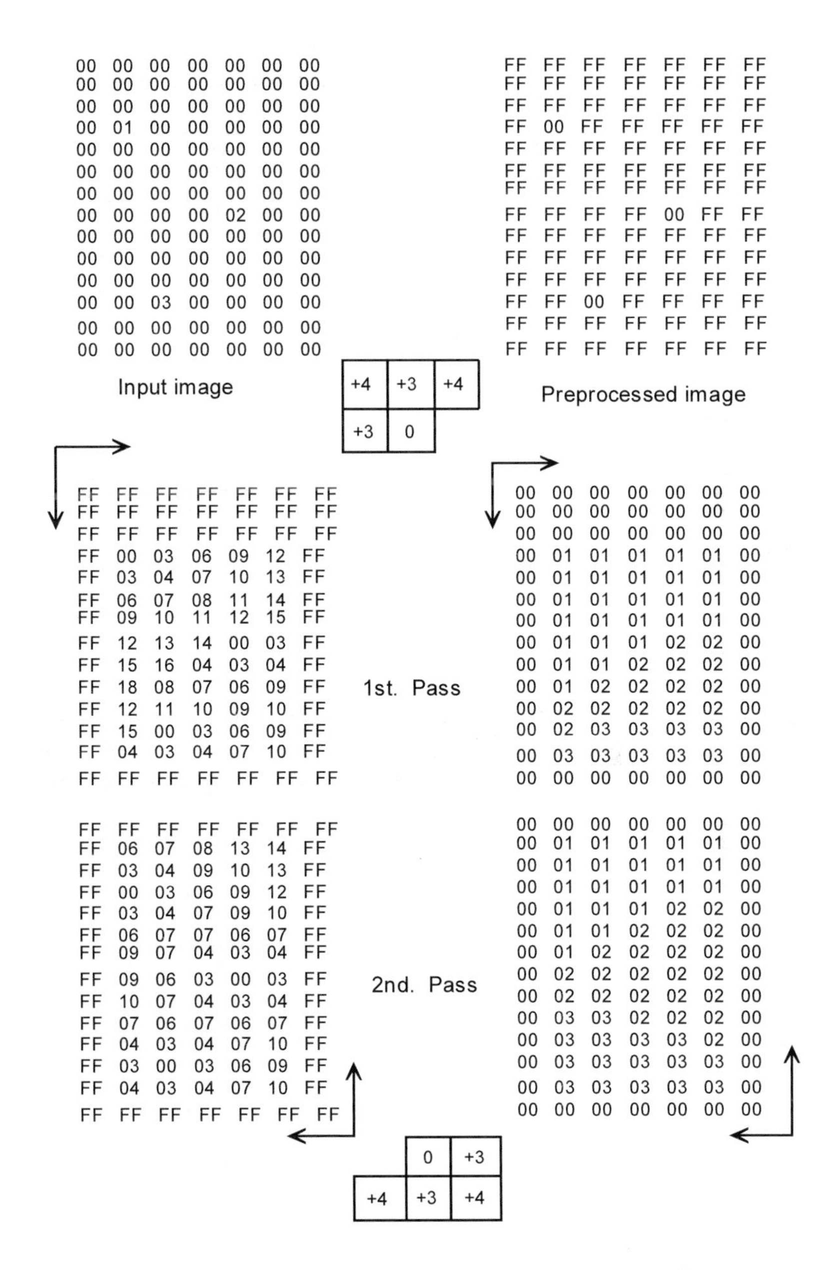

Fig. 8.6 DT and Voronoi tessellation parallel computation

The algorithm is tested by using several simulated digitized datasets (Figure 8.7) as well as photogrammetrically captured datasets (Figure 8.9).

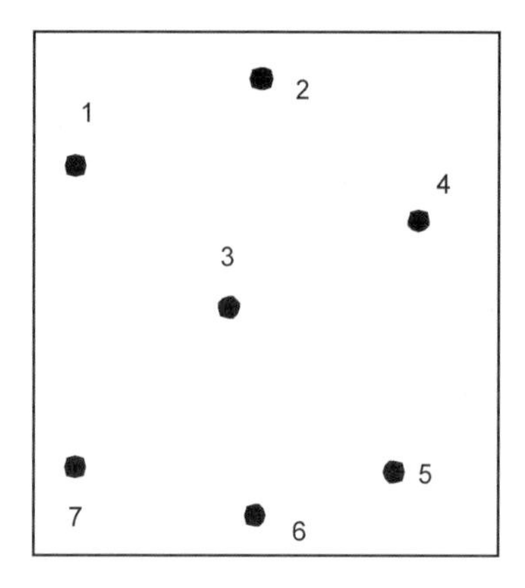

Fig. 8.7 Several kernel points

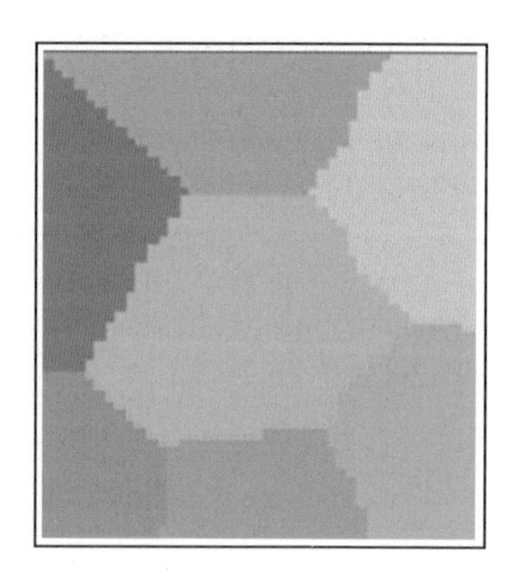

Fig. 8.8 The generated Voronoi polygons of the points as shown in Fig. 8.7

The Voronoi polygons in Figure 8.8 are clearly delineated. Different image tones represent different polygons as depicted in Figure 8.10 where their kernel points are shown in Figure 8.9.

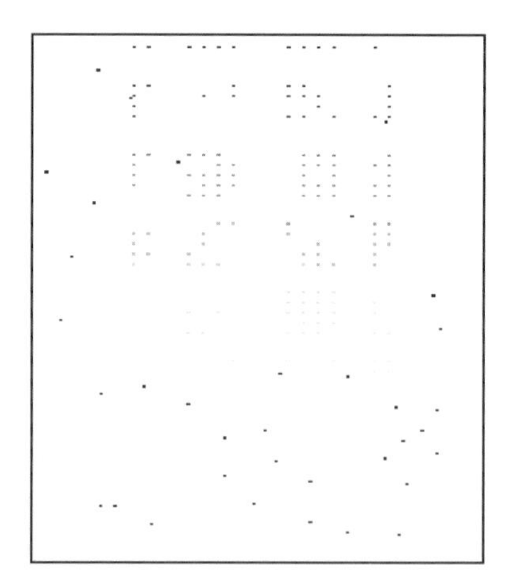

Fig. 8.9 The rasterized kernel points of the photogrammetric data sets

Fig. 8.10 The generated Voronoi polygons of the kernel points (Fig. 8.9)

8.4 Triangulations (TINs)

Descriptions of triangulations associated with digital terrain modelling (DTM) and surveying can be found in texts such as Petrie and Kennie (1990). A more specific discussion of TIN algorithms for visualization aspect can be found in (van Kreveld, 1997). In this section, the basic Delaunay triangulation (Delaunay, 1934) will be described.

The principle of Delaunay triangulation is that the circumscribing circle of any triangle does not contain any point of the data set inside it, see Figure 8.11.

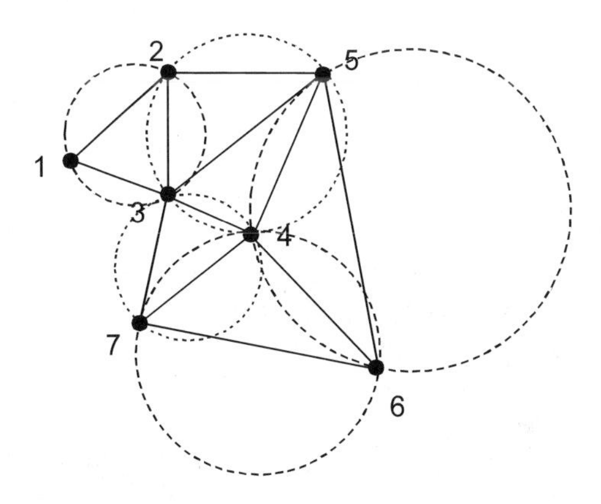

Fig. 8.11 Six non-overlapped triangles of seven points created by the Delaunay triangulation technique

A number of triangulation algorithms have been developed based on Delaunay triangulations and are widely implemented in terrain surface modules in a number of GIS and DTM packages. In such packages, the triangulation is normally known as a triangular irregular network (TIN). Each triangle in a TIN connects three neighbouring points so that the plane of the triangle fits the surface sufficiently. The TIN structure was designed by Peucker and co-workers (Peucker *et al.,* 1978) for digital terrain modelling. As mentioned in the foregoing discussion, a TIN is a terrain model that uses a sheet of continuous, connected facets based on a Delaunay triangulation of irregularly spaced nodes or observation points. TIN is considered to provide a better structure for surface modelling than other structure such as grids (grid for example may not retain the original data). It is not the intention of this section to describe fully the advantages of the structure but rather to briefly mention the TIN primitives instead. The primitives are nodes, lines, and surfaces which are considered the fundamental building blocks for spatial information. This is an interesting consideration from which to develop and implement the TIN package discussed in this work. In two-dimensional space, the 2D TIN can be used for developing a spatial information system; this is because the structure contains spatial data primitives, namely node, line, and surface primitives.

At this point unconstrained triangulations have been developed; that is no other terrain features incorporated such as breaklines or any linear features except terrain points. A much better triangulation, that is constrained triangulation capable of incorporating such terrain features is discussed in section 8.9. Three kernel points of the neighbouring Voronoi polygons need to be known to form a triangle as shown in Figure 8.12. If there are more than three neighbouring polygons, for example if there four polygons, then there will be two possibilities for triangle formation (see Figure 8.12).

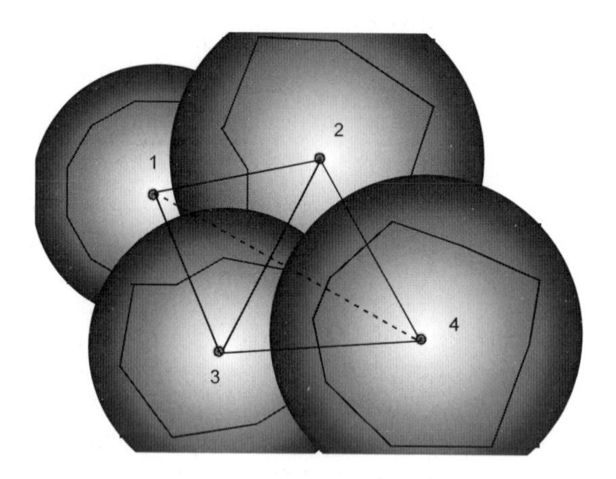

Fig. 8.12 The two possible triangles formation

In this work, the triangles are properly constructed according to the Delaunay technique where there are no ambiguous triangles created.

In other words, a correct TIN topology is established. A correct triangle formation can be achieved by searching three Voronoi polygon neighbours. In order to find a unique set of three points from a Voronoi-tessellated image, a 2 x 2 mask is used (as illustrated in Figure 8.13). The mask is designed to detect only two specific situations where three or four different pixel values fall inside the mask at a time. These different pixels correspond to the neighbouring Voronoi polygons and the kernel points of these polygons are used to form the triangle. Figure 8.13 shows the mask for detecting the triangle topology.

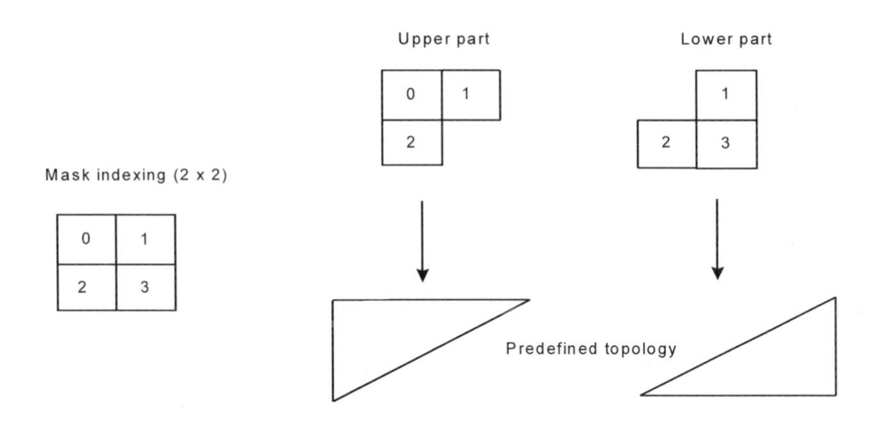

Fig. 8.13 Mask (2 x 2) for TIN topology detection

The mask is separated into two parts with the aim of avoiding the overlapping (crossover) triangles, as overlapping triangles are not allowed in the Delaunay triangulation. The mask (2 x 2) is designed to work using a matching operation. The pseudo code for the upper-part mask as follows:

```
if      (mask[0] not equal to mask[1]) and
          (mask[1] not equal to mask[2]) and
          (mask[2] not equal to mask[0]) then
          {
                    increase( number of triangles);
                    node[0] = mask[0];
                    node[1] = mask[1];
                    node[2] = mask[2];
}
```

whereas below is the matching condition for the lower-part of the mask:

```
if      (mask[1] not equal to mask[2]) and
          (mask[2] not equal to mask[3]) and
```

```
(mask[1] not equal to mask[3]) then
{
        increase(number of triangles)
        node[0] = mask[1];
        node[1] = mask[2];
        node[2] = mask[3];
}
```

The triangle detection also works with two passes of operations as for the previously discussed DT and Voronoi tessellation operations. The upper-part mask is used to scan the Voronoi image from the first pixel to the last pixel. A triangle is found if four different pixels match either one of the matching conditions imposed by the mask (see Figure 8.14). The figure illustrates how triangles could be detected.

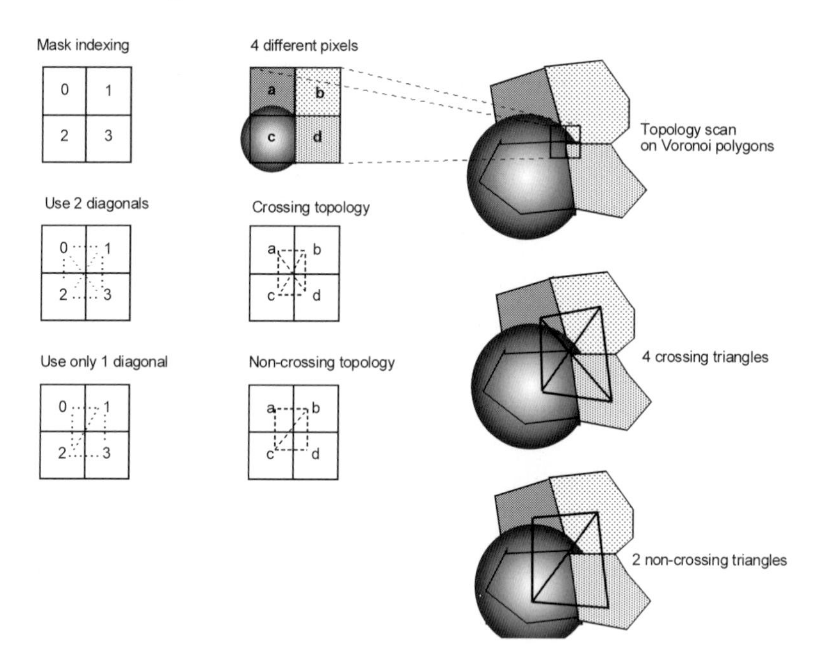

Fig. 8.14 Triangle topology detection

The above topology matching condition works only if non-adjacent rasterised points are found in the data set. In other words, two adjacent pixels of rasterised points produce incorrect topology (i.e. a very narrow polygon creates crossing triangles). This situation can happen if one chooses an inappropriate pixel size at the rasterising stage of the data sets.

Thus to overcome this problem, a few lines of condition codes are added to the previous matching conditions. The matching conditions are as follows:

```
if      (mask[0] not equal to mask[3]) and
if      (mask[0] not equal to mask[1]) and
           (mask[1] not equal to mask[2]) and
           (mask[2] not equal to mask[0]) then
           {
                   increase( number of triangles);
                   node[0] = mask[0];
                   node[1] = mask[1];
                   node[2] = mask[2];
                   Add triangle to the list;
           }

if      (mask[0] not equal to mask[3]) and
if      (mask[1] not equal to mask[2]) and
           (mask[2] not equal to mask[3]) and
           (mask[1] not equal to mask[3]) then
           {
                   increase(number of triangles)
                   node[0] = mask[1];
                   node[1] = mask[2];
                   node[2] = mask[3];
                   Add triangle to the list;
           }
```

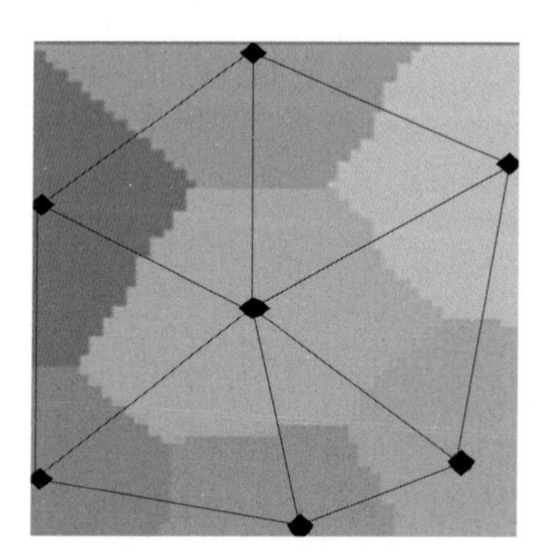

Fig. 8.15 The Voronoi polygons and its dual product (i.e. the triangles

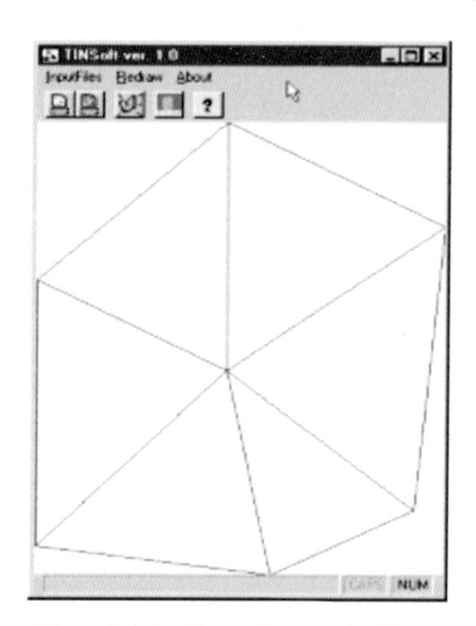

Fig. 8.16 The detected TINs from the Voronoi tessellated image

The triangle detection algorithm implementation works. Figure 8.15 and Figure 8.16 indicate the workability of the algorithm.

8.4.1 TIN Topological Data Structuring

A program has been developed for establishing TIN neighbour information (i.e. TIN topology). With this, one could determine the neighbours (neighbouring triangles) of any given triangles. This is very useful for some applications using the TIN data structure. The algorithm to establish the neighbour triangles is based on the following concept: a triangle neighbour is found if two common nodes of the triangles are encountered. One triangle may have a maximum of three different neighbours. Below is the pseudo-code for the algorithm.

```
loop (from triangle(t) = 1 to last)
      {
      loop (from triangle(tt) = 1 to last && num of neighbour <= 3)
        {
          if (t == tt) continue;

          set CommonNode = 0;

          loop (from node = 0 to < 3) && (CommonNode <= 2)
```

```
                  {
CheckNode =          (tri[t] -> Node[i] == tri[tt] -> Node[0]) ||
                     (tri[t] -> Node[i] == tri[tt] -> Node[1]) ||
                     (tri[t] -> Node[i] == tri[tt] -> Node[2]);

      if (CheckNode == true)
        {
        CommonNode ++;     // increase the common node
        if (CommonNode == 2)
          {
          NumofNbr ++;     // increase the number of
                              neighbour

           Nbr[NumofNbr] = tt;        // this triangle
           TotalNeighbour = NumofNbr + 1;// set total
                                 neighbours for a triangle
        }
      }
    }
  }
}
```

The input is a TIN file (an ascii file of three triangle nodes; Node1, Node2, Node3), and the output is an NBR file (a file of triangle number, number of neighbour, Neighbour[1] or (Nbr1), Neighbour[2] or (Nbr2), and Neighbour[3] or (Nbr3)), see Figure 8.17. The links of the dotted circles show that the triangle T1 and the triangle T2 are neighbouring triangles.

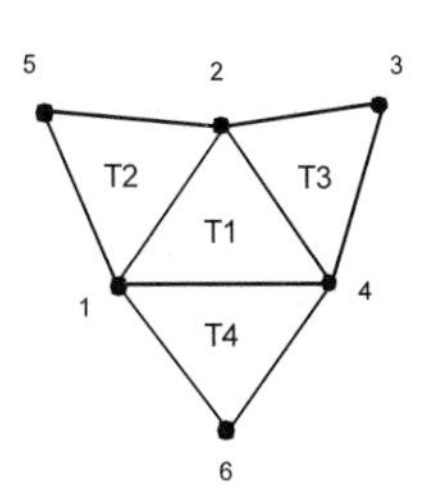

TIN file

#	Node1	Node2	Node3
T1	1	2	4
T2	1	2	5
T3	2	3	4
T4	1	4	6

NBR file

#	Num_Nbrs	Nbr1	Nbr2	Nbr3
T1	3	T2	T3	T4
T2	1	T1	-	-
T3	1	T1	-	-
T4	1	T1	-	-

Fig. 8.17 The TIN neighbour data structure

Full neighbouring information for the triangles is well described in the NBR file, and the link of the

XYZ coordinates with the TIN file (Figure 8.18) facilitates other tasks such as visualization.

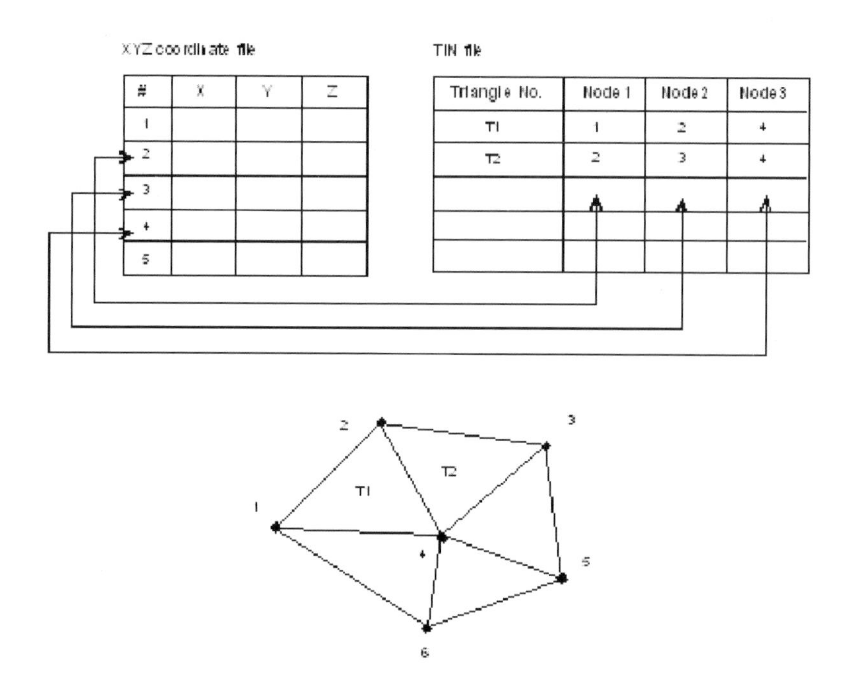

Fig. 8.18 The link of XYZ coordinates and the TINs

8.5 Visualization

It has been claimed by de Berg (1997) that the visualization of TINs is one of the major issues in TIN development. In this work only a simple display program for visualizing the generated TINs has been developed. One of the fundamental tasks of any GIS or DTM package is to visualize data. Figure 8.20 shows a simple TIN visualization.

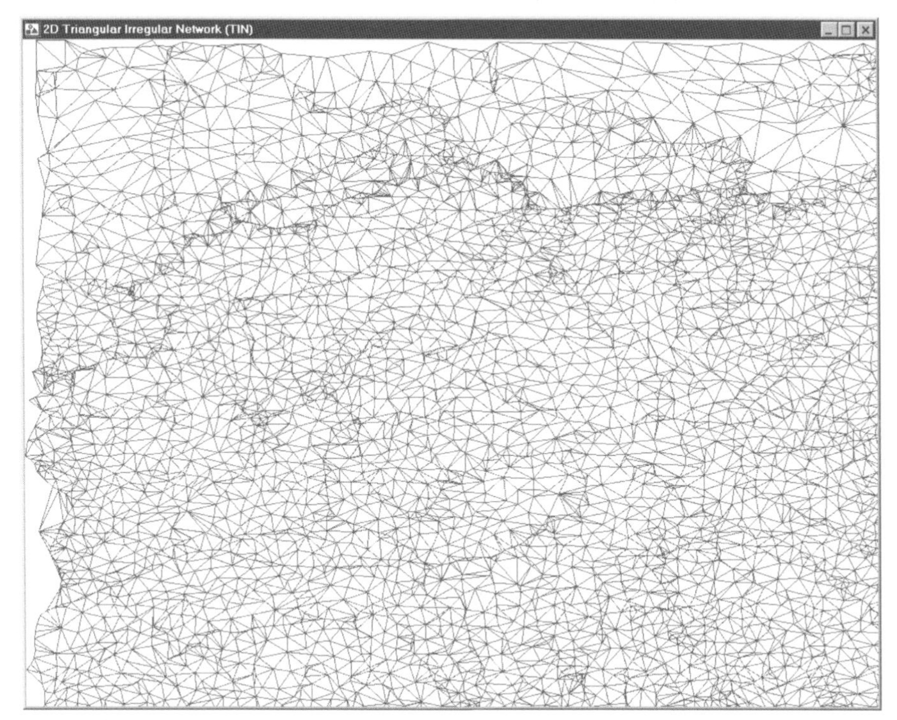

Fig. 8.19 The visualization of TINs generated using digitized contours data sets

The visualization program takes two input files, a XYZ coordinate file, and the TIN table file. The triangles three nodes (i.e. Node1, Node2, and Node3) can be linked to the corresponding XYZ coordinate table for the nodes with the appropriate pointers. Based on values in the XYZ file, triangles could be shaded according to slope, elevation, etc., for further visualization.

8.6 3D Distance Transformation

Digital distance transformations in 3D have been considered for more than a decade, not only in medical imaging but also in other areas (Borgefors, 1996). In this work, the DT technique was used to generate a DT image, a Voronoi image and tetrahedra. The 2D DT algorithm discussed in previous sections can be extended to the third dimension relatively straightforwardly due to the nature of the raster data structure. Thus, the same DT principle is utilised for the 3D TIN development. A 3D mask of dimension 3 x 3 x 3 was used as proposed by Borgefors (1996) known as Chamfer 3-4-5 mask (see Figure 8.20). Other types of masks are also applicable such as the Chessboard mask and the City-block mask (Borgefors, 1996).

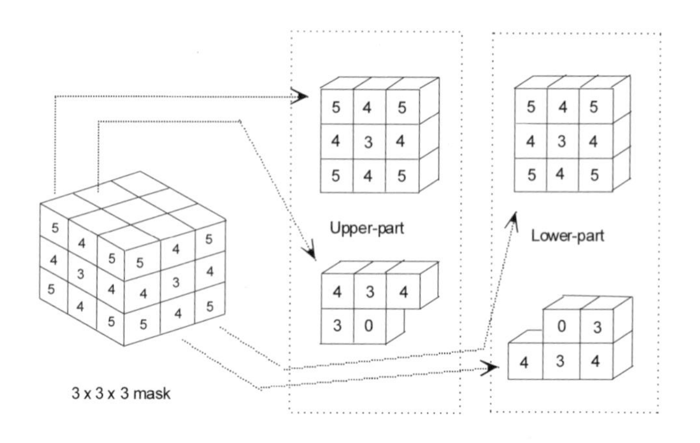

Fig. 8.20 The 3-4-5 mask for the 3D DT

The Chamfer 3-4-5 mask is used due to its computational simplicity and its ability to generate quite accurate distance images. Each voxel in the mask is assigned a local distance either with a value 3, 4 or 5, depending on the voxel location (again, see Figure 8.20). The centre voxel of the mask is surrounded by 26 other voxels in x, y, z directions, where each voxel has three types of voxel neighbours. They are called face neighbours, edge neighbours and node or vertex neighbours. The face neighbour voxels are assigned the value 3, the edge voxels the value 4, and the vertex voxels the value 5.

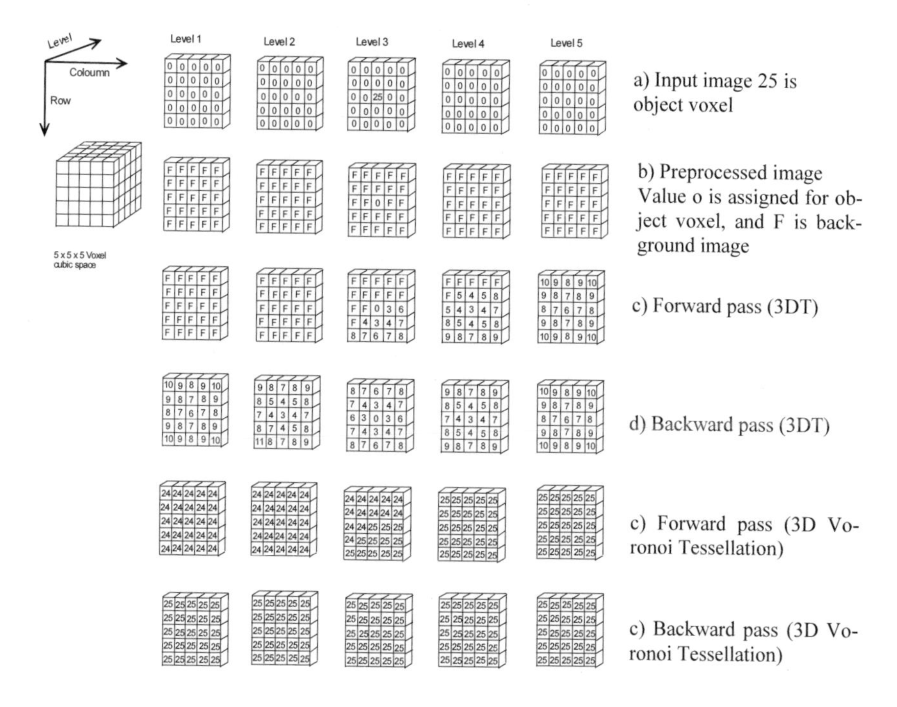

Fig. 8.21 Slice of images (along the Z or level direction) for the 3D DT and 3D Voronoi tessellation

Figure 8.21 shows how the voxel values are accumulated within a (5 x 5 x 5) voxel space in the DT and Voronoi operations. To generate a distance image of a 3D raster image, the first step is to set the voxel background image to the highest integer value (F) and the object voxels to zeros (i.e. 0), see b. The image is then scanned in two passes, i.e. forward and backward passes. The forward pass (using the upper-part mask) begins from the first voxel to the last voxel. At this stage, the voxels surrounding the object voxels will get new values. The new value is the minimum distance from the 14 possible voxel candidates (see c). The result of the first pass is taken into account for the second pass. This time, the image is scanned with the lower-part mask (i.e. the backward pass) beginning from the last voxel and moving to the first voxel; again see Figure 8.21 for the accumulated distance of a 5 x 5 x 5 cubic space (see d). A 3D distance-transformed image is formed after the two passes are carried out (see Figure 8.22). The Figure 8.22 shows the graphic output of the 3D DT of several random points in 3D space.

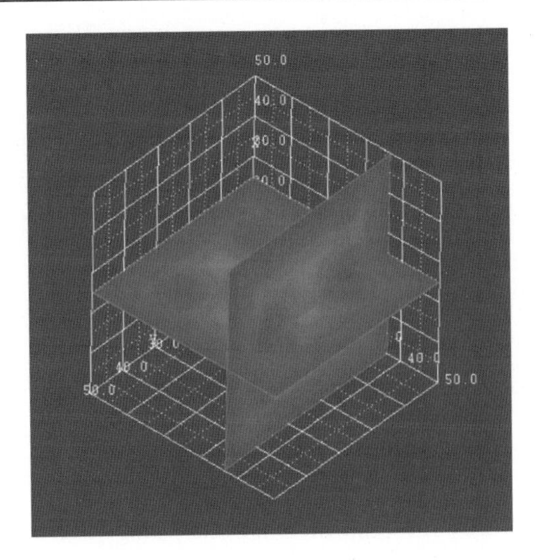

Fig. 8.22 An example of a 3D distance transformation image of four points shown as double cross-sections of a 3D space (visualized via the AVS software) in the x, y and x, z planes

The algorithm for the 3D DT in a pseudo-code follows:

```
// Procedure to set the background image
void set background()
  {
  loop from first row to last row
    {
     loop from first column to last column
       {
Set loop for (first level to last level
       {
            if (Pixel value not equal to background)



       Set Pixel value to zero;
            else
          Set Pixel value to background (highest possible value);
          }
       }
    }
  }
```

```
// Procedure to assign the Upper Mask
void GetUpperMask()
  {
   Assign the Mask[0] to Mask[4] to their corresponding pixel
locations,
    e.g. Mask[0] = Pixel[row-1][column-1];

// Procedure to assign the Lower Mask
void GetLowerMask()
  {
   Assign the Mask[4] to Mask[8] to their corresponding pixel
locations,
    e.g. Mask[4] = Pixel[row][column];
  }

// Procedure to compute distance in forward pass
void ForwardPass()
  {
   loop from first row to last row
    {
     loop from first col to last col
      {
       GetUpperMask();
       If Mask has odd index add 4 to the Mask value;
       else
       Add 3 to the Mask value;
        }
       Get the minimum value of Mask[0] to Mask[4] and assign
       to this pixel;
      }
    }
  }

// Procedure to compute distance in backward pass
void BackwardPass()
  {
   loop from last row to first row
    {
     loop from last column to first column
      {
       GetLowerMask();
```

```
            If Mask has odd index add 4 to the Mask value;
            else
            Add 3 to the Mask value;
             }
            Get the minimum value of Mask[0] to Mask[4] and assign
            to this pixel;
           }
          }
        }
```

and finally we need to combine the above steps into the following step:

```
     // Procedure to compute the distance using forward and back-
    ward
     void Forward&Backward()
      {
       Reads the input Image;
       Set the Background;
       Compute distance using the ForwardPass;
       Compute distance using the BackwardPass;
       Write and save the transformed image to file;
      }
```

8.7 3D Voronoi Tessellation

A Voronoi image is generated from the DT image. Again, these two images are generated in parallel. The task also involves three steps. First, cover the image with the mask. Second, the values of the mask are added to the value of the voxels being covered by the mask. Third, a minimum value from the 14 voxel candidates is determined and assigned to the current voxel location. The original voxel value of the current voxel location is taken, assigned, and written to the 3D Voronoi file. This is done prior to the mask being moved to the next voxel location. The process continues until the last voxel of the image is reached. Again, the result of this forward pass is taken into account in the backward pass which begins from the last voxel and proceeds to the first voxel of the image. Figure 8.21 (e and f) shows how the 3D Voronoi polygons (i.e. polyhedrons) are generated from one object voxel with ID = 25. In other words, a polyhedron of the voxels with ID 25 has been created. Visualization of the 3D DT and

3D Voronoi images or polyhedrons can be achieved by a true 3D viewing package as provided by the AVS™ software (see Figure 8.23).

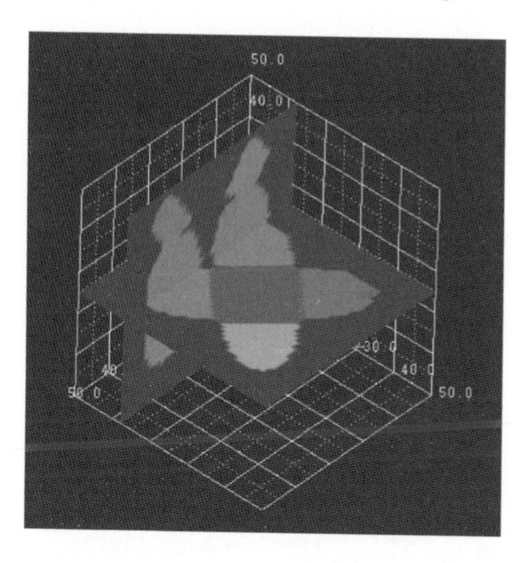

Fig. 8.23 An example of 3D Voronoi tessellation of four points shown as double cross-section of 3D space (visualized via the AVS software)

The algorithm for the above 3D Voronoi tessellation in pseudo-code can be written as:

```
// Procedure: SetBackground
void SetBackground(Voxel3D Voxel, unsigned char Bg, unsigned char Fg)
{
loop from first level to last level
  {
  loop from first row to last row
    {
    loop from first column to last column
      {
      if (Voxel[l][row][col] == 0)
         Voxel[l][row][col] = Bg;
      else
      if (Fg > 0)
         Voxel[l][row][col] = Fg;
      }
```

```
                }
        }
    }

// Procedure: GetUpperMaskDist
void GetUpperMaskDist(int l, int r, int c, Voxel3D Voxel,
Mask& MaskPix)
    {
    Assign the MaskPix[0] to MaskPix[13] to their corresponding
Voxel locations.
        e.g. MaskPix[0] = Voxel[l-1][r-1][c-1];
 }

// Procedure: GetLowerMaskDist
void GetLowerMaskDist(int l, int r, int c, Voxel3D Voxel,
Mask& MaskPix)
    {
    Assign the MaskPix[13] to MaskPix[26] to their correspond-
ing Voxel locations.
        e.g. MaskPix[13] = Voxel[l][r][c];
    }

// Procedure: GetUpperMaskVoronoi
void GetUpperMaskVoronoi(int l, int r, int c, Voxel3D Voxel-
Vor, Mask& MaskPixVor)
    {
    Assign the MaskPixVor[0] to MaskPixVox[13] to their corre-
sponding Voxel locations,
        e.g. MaskPixVox[0] = VoxelVor[l-1][r-1][c];
    }

// Procedure: GetLowerMaskVoronoi
void GetLowerMaskVoronoi(int l, int r, int c, Voxel3D Voxel-
Vor, Mask& MaskPixVor)
    {
    Assign the MaskPixVor[13] to MaskPixVox[26] to their cor-
responding Voxel locations,
        e.g. MaskPixVox[13] = VoxelVor[l][r][c];
    }
```

```
// Procedure: ForwardPass
void ForwardPass(Voxel3D Voxel, Voxel3D VoxelVor)
 {
 loop from first level to last level
   {
   loop from first row to last row
     {
     loop from first column to last column
       {
       GetUpperMaskDist(l, r, c, Voxel, MaskPix);
       GetUpperMaskVoronoi(l, r, c, VoxelVor, MaskPixVor);
       for (k = 0; k < 13; k ++)
         {
         if ((k == 0) || (k == 2) ||
            (k == 6) || (k == 8))
           MaskPix[k] = MaskPix[k] + 5;

         if ((k == 1) || (k == 3) ||
            (k == 5) || (k == 7) ||
            (k == 9) || (k == 11))
           MaskPix[k] = MaskPix[k] + 4;

         if ((k == 4) || (k == 10) || (k == 12))
           MaskPix[k] = MaskPix[k] + 3;
         }

       if (MaskPix[13] != 255)
         MaskPix[13] = 0;

       Voxel[l][r][c] = MaskPix[MinByIndex(0, 13)];
       VoxelVor[l][r][c] = MaskPixVor[MinByIndex(0, 13)];
       }
     }
   }
}
```

```
// Procedure: BackwardPass
void BackwardPass(Voxel3D Voxel, Voxel3D VoxelVor)
 {
 loop from last level to first level
   {
   loop from last row to first row
     {
     loop from last column to first column
       {
       GetLowerMaskDist(l, r, c, Voxel, MaskPix);
       GetLowerMaskVoronoi(l, r, c, VoxelVor, MaskPix-
Vor);
       for (k = 26; k > 13; k --)
         {
          if ((k == 18) || (k == 20) ||
             (k == 24) || (k == 26))
             MaskPix[k] = MaskPix[k] + 5;

          if ((k == 19) || (k == 21) ||
             (k == 23) || (k == 25) ||
             (k == 15) || (k == 17))
             MaskPix[k] = MaskPix[k] + 4;

          if ((k == 14) || (k == 16) || (k == 22))
             MaskPix[k] = MaskPix[k] + 3;
         }
        Voxel[l][r][c] = MaskPix[MinByIndex(13, 26)];
        VoxelVor[l][r][c] = MaskPixVor[MinByIndex(13, 26)];
        }
     }
   }
 }

// Procedure: ForwardVoronoi
void ForwardVoronoi()
 {
 ReadVoxelImage(Voxel);
 CopyVoxel(Voxel, VoxelVor);
 SetBackground(Voxel, 255, 0);
 ForwardPass(Voxel, VoxelVor);
 }
```

```
// Procedure: BackwardVoronoi
        void BackwardVoronoi()
        {
        BackwardPass(Voxel, VoxelVor);
        }
```

8.8 Tetrahedron Network (TEN) Generation

Using the same principle as for the 2D TIN, the algorithm for the 3D TIN utilises a mask of 2 x 2 x 2 (see Figure 8.24). It has eight voxel elements. It provides a unique way of establishing tetrahedra. In order to obtain non-overlapping tetrahedra, several predefined conditions have to be imposed during voxel scanning. There are six possible non-overlapping tetrahedra that we can get from the mask shown in Figure 8.24.

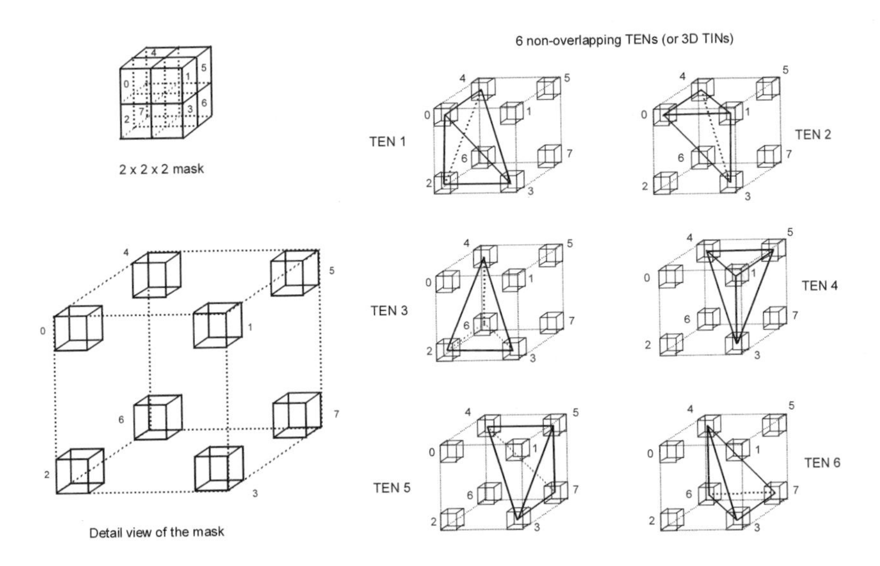

Fig. 8.24 The six non-overlapping TENs

The mask is then used to scan the voxel's Voronoi tessellated-image once. Once the tetrahedron is detected (based on the imposed conditions), it is then written to a file. The file contains a record of tetrahedra where each record has four nodes, it is an ASCII file and structured as in Figure 8.25. Thus, it is one way of establishing a simple tetrahedral data structure. The data structure together with a table of nodes' coordinates provide a means for further manipulation of the data, e.g. visualization.

The algorithm is implemented and tested by using simulated 3D raster data sets. This data sets are generated by the 3D point-to-raster program developed in this work. A wireframe display program has also been developed for visualizing the TENs (see Figure 8.26) for the output display.

Points table

#	X	Y	Z
0			
1			
2			
3			
4			

TENs table

#	Node1	Node2	Node3	Node4
1	0	2	3	4

Fig. 8.25 TEN data structure (for TEN 1 as shown in Fig. 8.24)

```
// The main program has the following routines:
        void main()
         {
          GetVPIfile(VPI);
          GetVPDfile();
          AllocateMemory();
          Get3DTINfile();
          Make3DTIN();
          DeallocateMemory();
         }
```

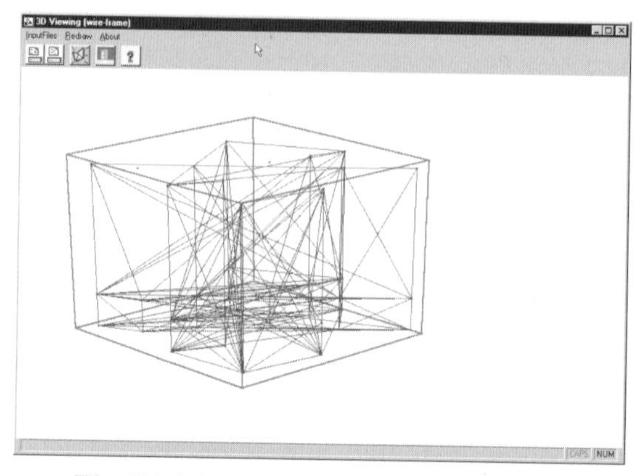

Fig. 8.26 An example of TEN visualization

8.9 Constrained Triangulations

Constrained triangulation development is meant to accommodate linear features, e.g. terrain breaklines, drainage lines, faults and other linear features such as roads, railways, etc. Previously, this triangulation only worked with points as discussed in section 8.4, but now a new feature is introduced in this work, that is the capability of handling linear features. In a constrained triangulation, these lines or linear features become part of the triangle edges. Since the triangulation in this work is based on raster data, further development of the rasterization routine has also been necessary so that it can accommodate straight lines.

The next section discusses the line rasterization algorithm and the results of constrained triangulation.

8.9.1 The Line Rasterization

Line rasterization is used to rasterise a series of lines - as input to the constrained triangulation. In this, a line has a start node and an end node. Thus, the rasterization is simply a process of calculating the position of the pixels between these two end nodes of a line. Line rasterization was successfully implemented for the constrained triangulation by incorporating a simple line equation, $y = mx + C$, where m is a slope of a line, the Bresenham line plotting algorithm, and the Tang (1992) algorithm. The m (slope of a line) is used only to detect the type of a line or arc in the data file rather than to do the rasterization of points of a line. A line or an arc could have a slope of $(0 < m < 1)$, or $(m > 1)$, or $(-1 < m < 0)$, or $(m < -1)$ or be vertical or it could be horizontal. This approach can handle all cases of lines, i.e. in all quadrants. The Bresenham algorithm is used to speed up the operation, and Tang's algorithm is for the constrained triangulation where the edge or arcs can be accommodated in the Voronoi and the triangulation process. This approach of line rasterization is fast and produces well rasterised lines. It is fast because it does not involve a divisional operation during the pixel increment along the line. The Bresenham line algorithm can be found described in computer graphics texts such as Foley, et al (1996), and Farrell (1994). Below is the line rasterization pseudo-code (only the first quadrant is presented).

```
void LineRasterization()
                    {
                  GetXYZFile();
                  GetArcFile();
                    loop from first arc to last arc
                         {
                         GetXYZforArcNodes(t, xstart, ystart, xend,
                         yend, sNd, eNd);
                       Calculate the m for each line (m = (yend -
              ystart) / (xend - xstart));
                         GetMiddleXY for each line;
                       Detect the slope of a line, it could be (m > 0)
                     && (m < 1), or
                           (m > 1), or (m < 0) && (m > -1), or (m < -
                         1), or


                           vertical, or horizontal line;
                    if (m > 0) && (m < 1) do the following
                       {
                       Swap the coodinates of the two nodes so that
                    the operation begins
                           with the lower position node;
                       Calculate the dx, dy;
                       Calculate the (2 * dx) and (2 * dy);
                       Calculate the pixel location, i.e. the row, and
                    col of the pixel location;
                       if swap the nodes is true
                           Assign the pixel location with the correct
                         value, i.e. the correct node number;
                       else
                           Assign the pixel location with the correct
                         value, i.e. the correct node number;
                       Initialise the Bresenham error of a line (that is
                    the difference of a point to the true position, see
                    Bresenham algorithm for detail.
                       Increase the xstart (i.e., xstart ++)
                       while (xstart < xend)
                           {
                              If the error > 0
                                   error = error + (2*dy - 2*dx);
                                   ystart ++;
```

```
            else
                  error = error + (2*dy);

            Calculate the row and col of the pixel
         location;
            Assign the correct pixel value to this
         location;
            Increase the xstart (i.e., xstart ++);
         }
and do the rest of the line cases (i.e. for
the other quadrants).
   }
```

This line rasterization algorithm produces the following result as tested with several simulated nodes, and arcs, as shown in Figure 8.27.

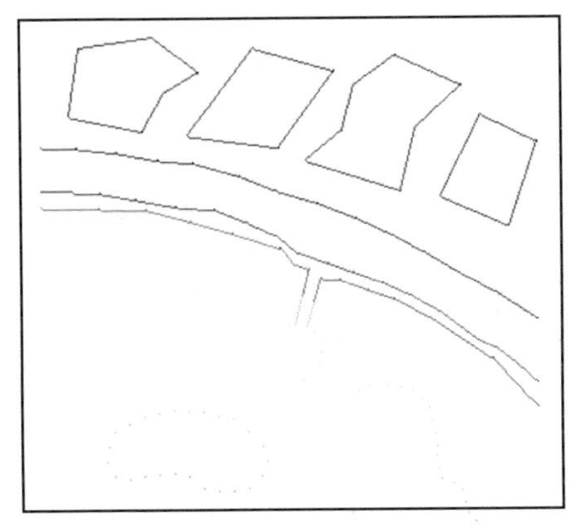

Fig. 8.27 Screen shot of the rasterized nodes and arcs.

8.9.2 The Construction of the Constrained TINs

In order to generate constrained triangulation, rasterised points and rasterised lines are needed as the input. This constrained triangulation is based on the concept presented by Tang (1992). Here, the constrained edges were represented by a series of pixels whose values were based on the edge node IDs (identifiers). In other words, half of a line was represented by the pixels of the start node, and the other half by the pixel values of the end node (see Figure 8.28).

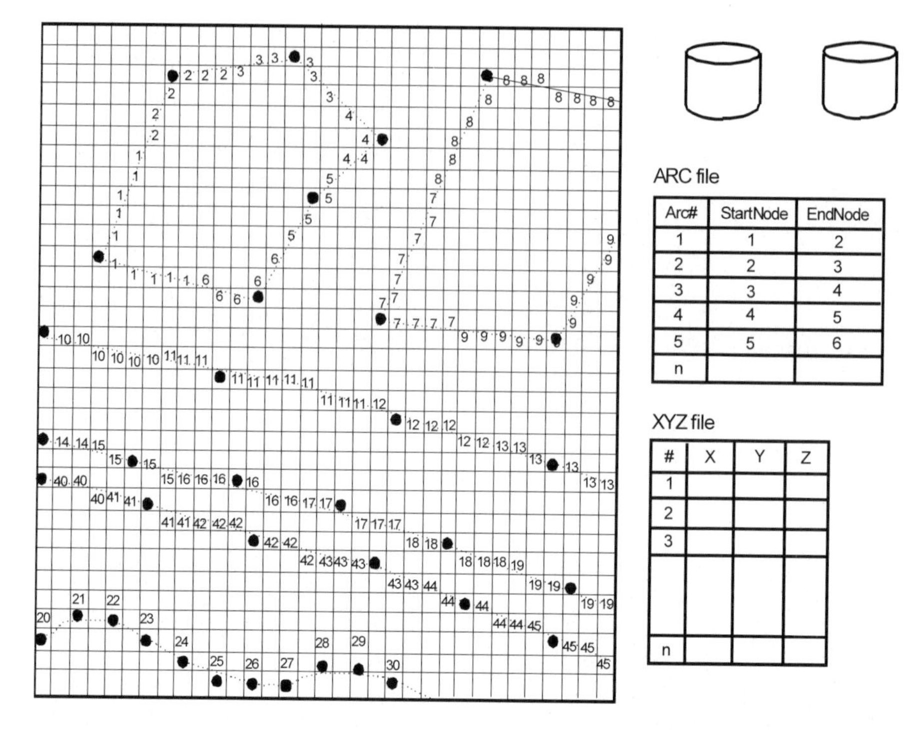

Arc#	StartNode	EndNode
1	1	2
2	2	3
3	3	4
4	4	5
5	5	6
n		

#	X	Y	Z
1			
2			
3			
n			

Fig. 8.28 An example of the pixel locations of the rasterized points and severals edges or arcs. The left side is the corresponding coordinates and arc files.

The above figure also shows the propagation of the pixel values of the start node to the end node of each arc. This approach also conforms with the Voronoi tessellation concept where two kernel points have two corresponding Voronoi zones separated by a boundary which happens to be located in the middle of the two kernel points.

The illustrations in Figure 8.29a to Figure 8.29d show the result of the DT and the Voronoi tessellation implemented for the constrained edges.

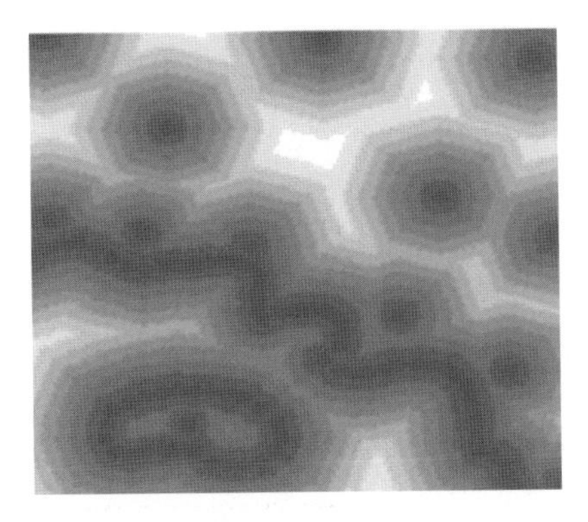

Fig. 8.29a The DT image of the rasterized kernel points

Fig. 8.29b The Voronoi image for the kernel points Fig. 8.29a

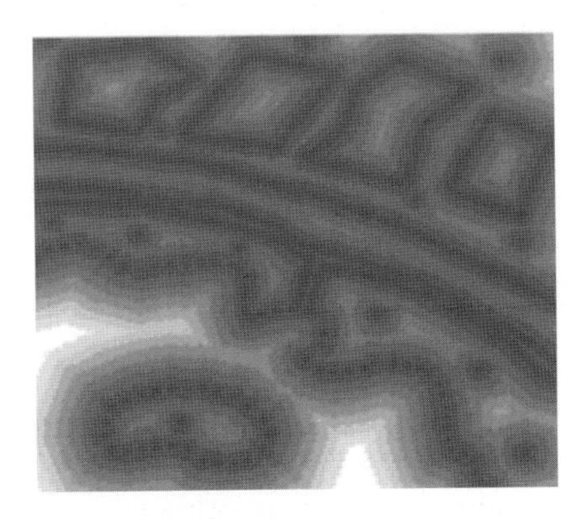

Fig. 8.29c The DT image with the edges and points

Fig. 8.29d The Voronoi image of the corresponding dges and points of Fig. 8.29c

It is clearly shown that the edges or arcs can be accommodated as a constrained feature in the distance transformation and Voronoi tessellation. A constrained edge is represented by the thick black lines as in Figure 8.29c and the respective polygons are shown in Figure 8.29d.

Further, all the points and the edges are then triangulated, and the results are the constrained triangulations, see Figure 8.31.

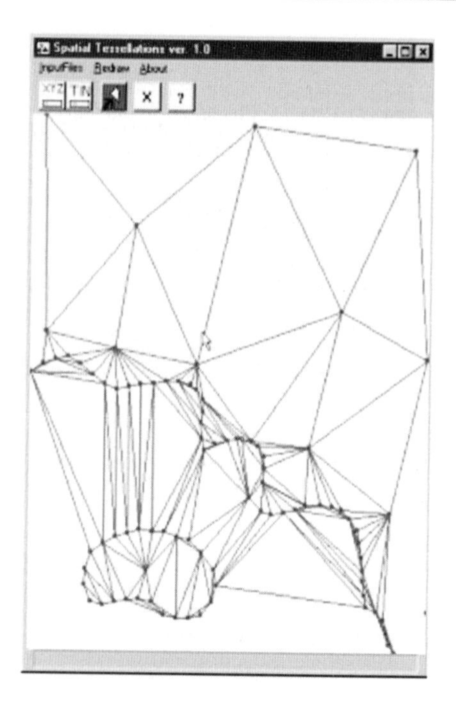

Fig. 8.30 The generated unconstrained triangulation

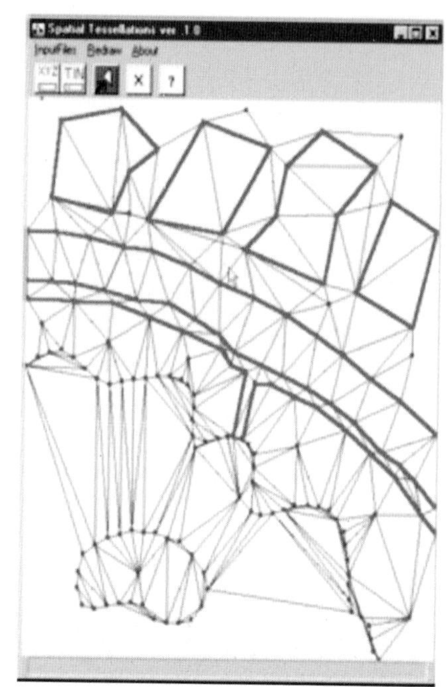

Fig. 8.31 The generated constrained triangulation

In this particular example, the edges are part of the features on the terrain and also form part of the triangle edges. The results indicate that the constrained triangulation works. The development provides useful data structuring mechanisms for TIN-based spatial data modelling and the related applications. The fundamental GIS data types, i.e. node, arc, surface and volume are generated with this approach. Their related spatial modelling is discussed in Chapter 7.

The technique has also been tested using photogrammetrically acquired data (Drumbuie, Kyle of Lochash, north-west Scotland) - see the results in the following Figure 8.32 to Figure 8.35.

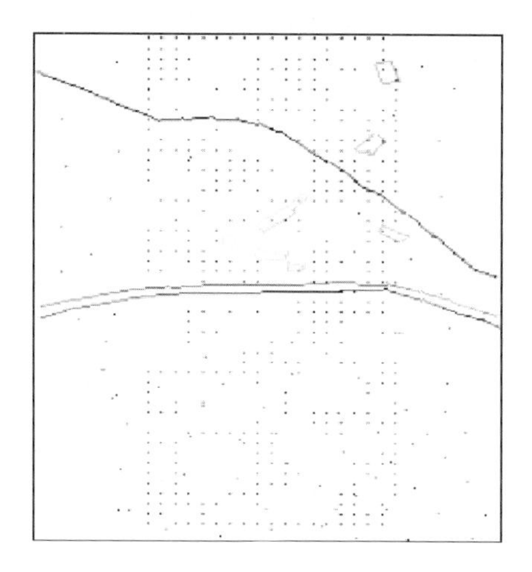

Fig. 8.32 The rasterized points and lines

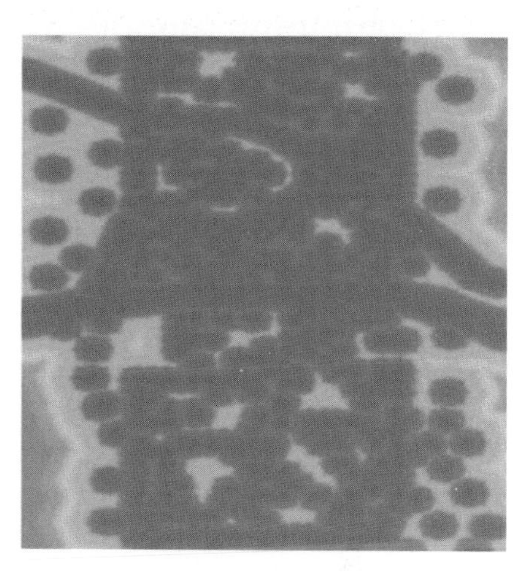

Fig. 8.33 The DT image of the area

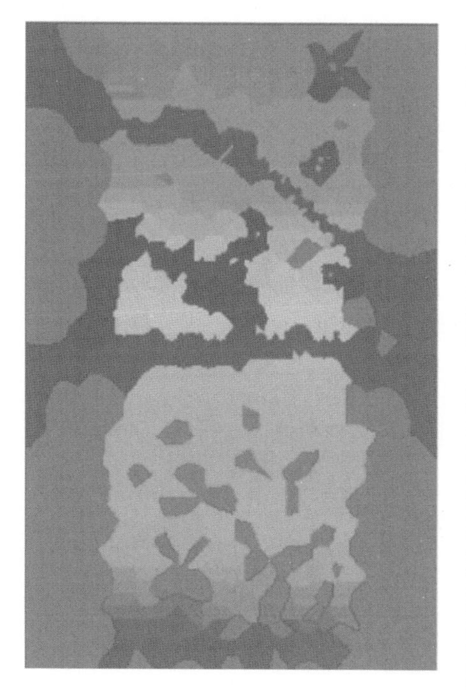

Fig. 8.34 The Voronoi image

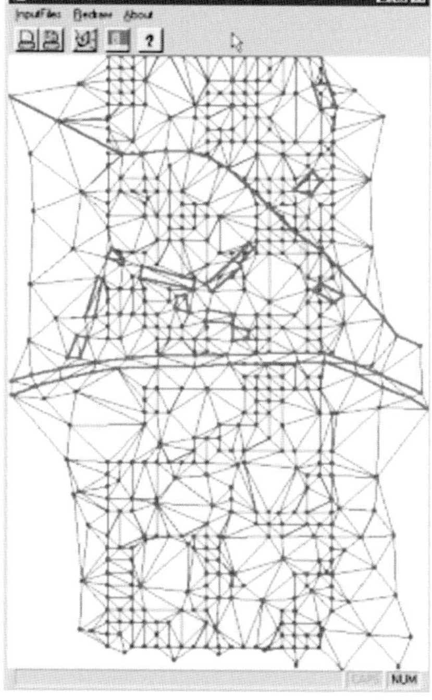

Fig. 8.35 The generated TINs

8.10 Contouring Algorithm

Contouring is one of the GIS applications that has been developed for this work. This section describes the development of the data structures for the contouring and contouring algorithm. A suitable file format for the above application is also developed, so that it could be imported to other commercial GIS software, e.g. Arc/Info, and ILWIS.

8.10.1 Data Structures for Contouring

This is one of the important components where the data storage and data's accessibility influence the behaviour and performance of the software. In this it should be noted, besides the two TINs structures, namely, the three-nodes table (TRI) and the triangle neighbour (NBR), that two other data structures were developed. These two TIN topological structures are a triangles' edge and right and left triangles (TRS), and a triangles' three sides (SID).

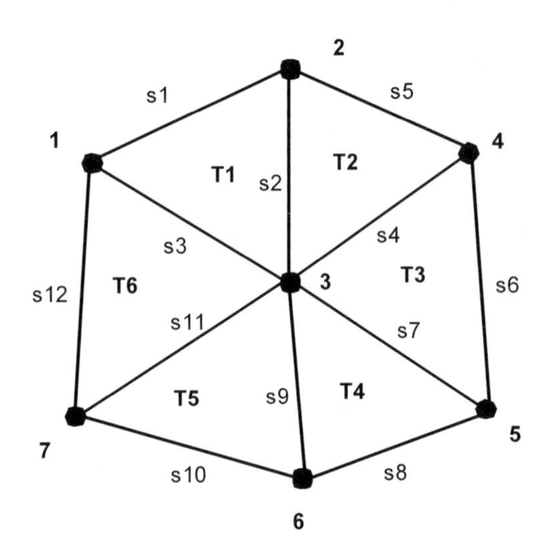

Fig. 8.36 An example of 6TINs with seven nodes, and 12 sides or edges

Figure 8.36 shows a simple configuration of six TINs with seven nodes, and 12 triangle sides.

To facilitate the contouring application, a program has been developed to generate two more data structures. Thus, from TRI (Triangle # and 3 nodes) and NBR (Triangle neighbours), TRS and SID structures are generated. The TRS contains Triangle sides and Right-Left triangles while the SID structure contains Triangle # and the three sides or edges.

The algorithm for converting the TRI and NBR structures to TRS and SID structures is based on the following concept: a triangle side has two nodes, and each side only has either a right triangle or a left triangle (see Figure 8.36). The software has the following routines:

Read the triangles
Read the triangles neighbours, and
MakeTRSandSID structure.

The MakeTRSandSID has the following sub procedures, the ExistingSide and the DoSide functions, and the algorithms are described below in pseudo-code.

```
ConvertStruct :: bool ExistingSides(int n1, int n2, int& s)
            {
            do
              {
              found = (n1 == Node1) && (n2 == Node2);
              if (! found)
                s ++;
              } while ((! found) && (s <= nsid));

             return found;
            }

           void ConvertStruct :: DoSide(int t, int n1, int n2, int
         snbr)
            {
            if (n1 > n2)
              {
              h = n1;
              n1 = n2;
              n2 = h;


              }
```

```
        if (ExistingSides(n1, n2, s))
         {
          TriSides[s]->RightTri = t;
          Tri3Sides[t]->Side[snbr] = s;
         }

        else
         {
          nsid ++;
          TriSides[nsid] = new tsid;
          TriSides[nsid]->Node1 = n1;
          TriSides[nsid]->Node2 = n2;
          TriSides[nsid]->LeftTri = t;
          TriSides[nsid]->RightTri = 0;

          Tri3Sides[t]->Side[snbr] = nsid;
         }
}
```

TRS structure

#	Node1	Node2	RightTri	LeftTri
s1	1	2	T1	0
s2	2	3	T1	T2
s3	1	3	T6	T1
s4	3	4	T3	T2
s5	2	4	T2	0
s6	4	5	T3	0
s7	3	5	T4	T3
s8	5	6	0	T4
s9	3	6	T9	T4
s10	6	7	T5	0

SID structure

#	Side1	Side2	Side3
T1	s1	s2	s3
T2	s2	s4	s5
T3	s4	s6	s7
T4	s7	s8	s9
T5	s9	s10	s11
T6	s3	s11	s12

Fig. 8.37 The TRS and SID structure

8.10.2 The Algorithm

The contouring program makes use of two TIN data structures, namely the TRS and SID structure plus the coordinates, and it is based on linear interpolation. The contouring program performs the following routines:

Read the input data (coordinates, TINs structure of TRS, and SID tables).

Open the output file for the interpolated data.

Get the min and max of the XYZ input coordinates, then perform

MakeContouring.

The algorithm can be described as follows. MakeContouring has several sub methods or procedures, they are CheckSide, FindFirstTri, Interpolate, FindOtherSide, FindNextTri, and GetContours. The CheckSide is to check the side of a triangle and whether or not it can be interpolated with the user requested contour heights. The FindFirstTri is used to get the first triangle which contains the requested contour height. FindOtherSide is to get the other side of a triangle, whereas the FindNextTri is to get the next triangle in the list. The Interpolate is to compute the interpolated point once a triangle's side satisfies the imposed conditions. Then all these sub procedures are combined as the GetContours function performs the subsequent major task - the contouring. Thus the GetContours behaves as follows.

```
void MakeContouring :: GetContours(int Hreq)
              {
                do
                 {
                 while (FindFirstTri(t, s) != 0)
                   {
                   ii ++;
                   SegNr = ii;
                   startt = t;
                   starts = s;
                   done = false;
                   secondpart = false;

                   do
                     {
                     Interpolate(s, SegNr, Hreq);
                     FindOtherSide(t, s, nexts);
                     FindNextTri(t, nexts, nextt);
                     if (nextt == startt)
                       {
                       // found closed contours
                              Interpolate(nexts, SegNr, Hreq);
                       done = true;

                       }
                     else if (nextt == 0)
                           {
                           // hit border
```

```
                    Interpolate(nexts, SegNr, Hreq);
                if (secondpart)
                  done = true;
                else
                  {
                  FindNextTri(startt, starts, t);
                  if (t == 0)
                     done = true;
                   else
                     {
                     s = starts;
                     secondpart = true;
                     }
                  }
                }
           else
             {
             t = nextt;
             s = nexts;
             }
          } while (! done);

       }

     } while ( ! ((t == 0) || (s == 0) || (z != prevH)) ) ;
}
```

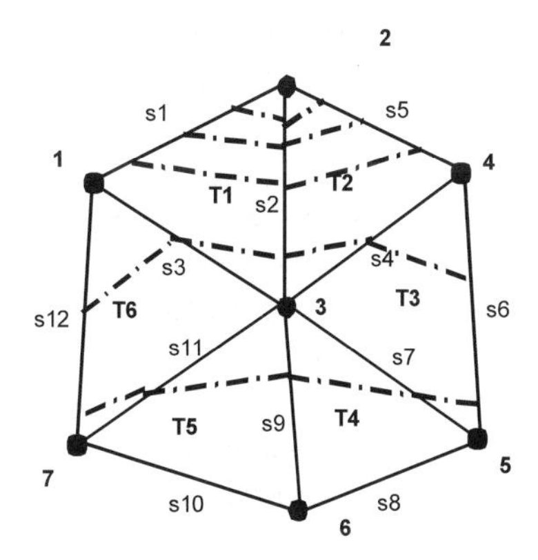

Fig. 8.38 An example of contours with 35 interval using six TINs with seven nodes, and 12 sides or edges

8.10.3 The Contour Visualization

The contour algorithm has been tested using real terrain data sets. Figure 8.39 and Figure 8.40 illustrate the generated contours from the simulated and digitized contours datasets with different contour intervals. Format conversion programs for contours display in other popular GIS packages are also developed, for example PC Arc/Info (.LIN format) and ILWIS packages (.SEG format). Figure 8.40 shows one of the examples of derived contours output from photogrammetrically acquired data sets.

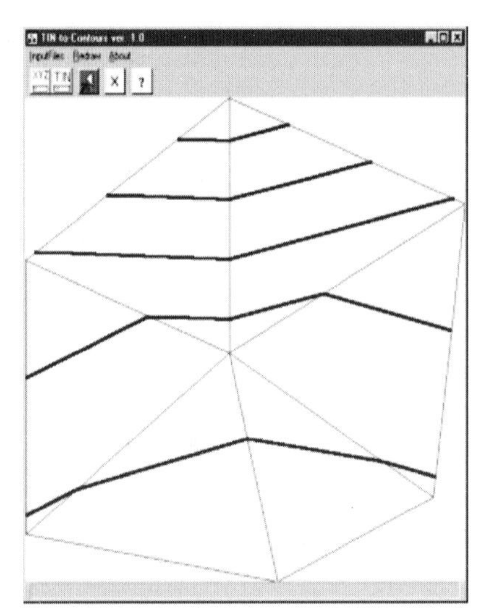

Fig. 8.39 The generated contours from the simulated datasets (6TINs)

Fig. 8.40 The generated contours of 4m interval using photogrammetrically datasets (Drumbuie, Kyle of Lochash, north-west Scotland)

8.11 Algorithms for Irregular Network Formation

For 2D simplicial network formation, scanning the Voronoi image using a 2x2 mask with predefined conditions generates the Delaunay triangulation. A triangle is detected in a situation where at least three of the four elements of the mask are different. By combinatorial mathematics (Finkbeiner and Lindstrom, 1987; Liu, 1986), this is a 3-selection (3-combination) from a mask, which is a set of at most four distinct elements. The 3-selection is a subset of the mask. We can apply the following definition to determine the number of combinations (triangles in this case):

The number of k-selections from an n-element set is denoted by C(n, k),

$$C(n,k) = \frac{n!}{(n-k)!\,k!}; \qquad \text{where } 0 \leq k \leq n$$

If the mask contains only three distinct pixels, it becomes:

$$C(3,3) = \frac{3!}{(3-3)!\,3!} = 1$$

which yields only one triangle.

If all four elements of the mask are different, it is:

$$C(4,3) = \frac{4!}{(4-3)!\,3!} = 4$$

which yields four triangles to be formed. This leads to four intersecting triangles. The situation occurs when 4 points are situated on a circle, hence, four Voronoi regions meet at the centre of this circle, as shown in Figure 8.41.

To overcome this problem, only the combinations from the two opposite diagonals of the mask are selected, which yield only two non-intersecting triangles.

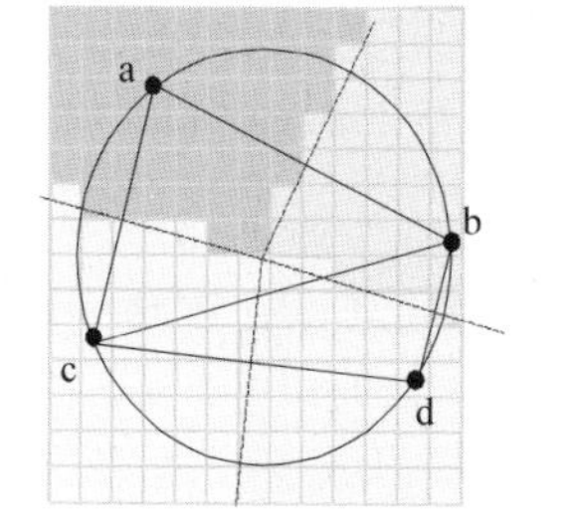

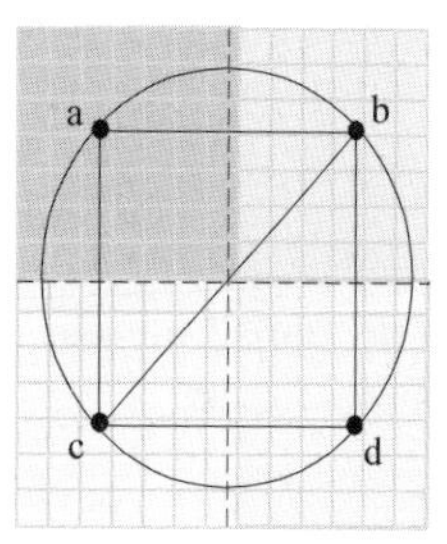

Fig. 8.41 Four points situation a circle four Voronoi regions meet ay its centre

The above analysis leads to the conditions i) and ii) for triangulation. Let a, b, c and d be the contents of the 2x2 mask at an instance:

Condition i) The elements in the upper triangle of the mask must be different (elements number 1, 2, 3 in Figure 8.42). This allows four possibilities:

a b	a b a b	a b
c c	c b c d	c a

or

Condition ii) The elements in the lower triangle of the mask must be different (elements numbered 2, 3, 4 in Figure 8.42). This allows four possibilities

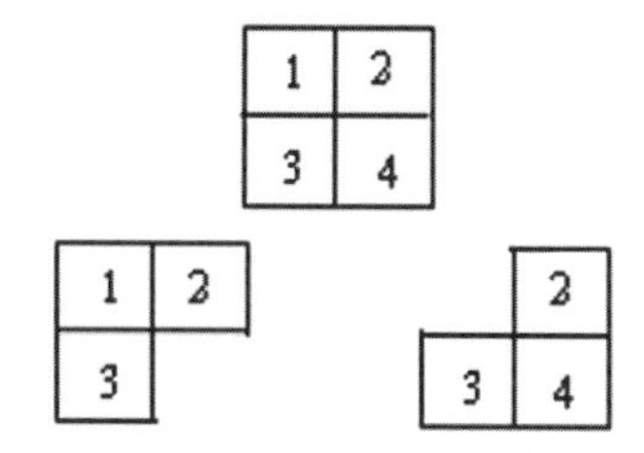

Fig. 8.42 2x2 mask elements numbering

a a a b a b a b

b c a c c d c a

Condition iii) The two elements on the perpendicular diagonal must be different (elements numbered 1, 4 in Figure 8.42). This condition prevents a faulty formation of a triangle caused by a sliver raster Thiessen polygon. A sliver polygon may be caused by a situation as shown in Figure 8.43.

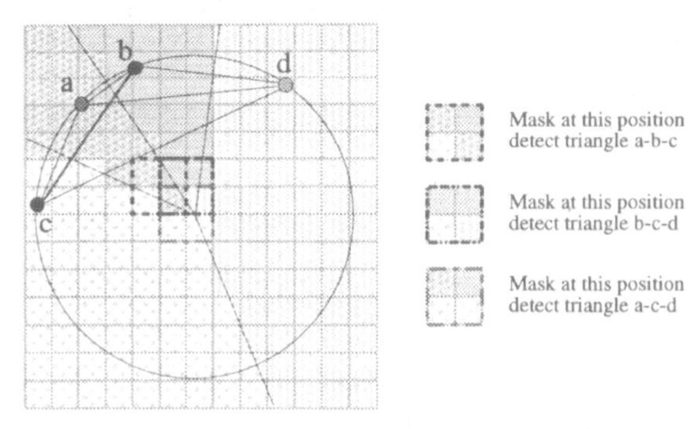

Fig. 8.43 Faulty triangulation caused by sliver Voronoi region

The consequence of this condition is the rejection of the following situation:

a b

c a

The union of the permissible situations in condition i) and ii) yields six distinct possibilities. One of these is rejected by condition iii), so that only five possibilities remain.

a b a b a a a b a b

c c c b b c a c c d

For 3D, the formation of a 3D simplicial network is done by using the 2x2x2 mask to scan the 3D Voronoi array once. A tetrahedron is detected when the contents of four elements of the mask from the total of eight are different. By combinatorial mathematics, this is a 4-selection from a set of eight distinct voxels, that is:

$$C(8,4) = \frac{8!}{(8-4)!\;3!} = 70$$

This means that there are 70 possible intersecting tetrahedrons.

To ensure proper formation of the 3D simplicial network, that is, the network of tetrahedrons, a Boolean approach is used to set up a set of conditions to form tetrahedrons. The general aim is to allow the creation of at most six non-intersecting tetrahedrons at one instance if all eight elements of the mask are different (see Figure 8.44).

To achieve this requirement, six primary conditions are attached to the 2x2x2 mask for the formation of a unique set of non-overlapping tetrahedrons. Note that the numbers encompassed by circles in Figure 8.44 correspond to the following conditions (1) to (6). The numbering system of the mask is shown by the numbers encompassed by small cubes.

(1) $1 \neq 3 \neq 4 \neq 5$ (2) $1 \neq 2 \neq 4 \neq 5$ (3) $3 \neq 4 \neq 5 \neq 7$

(4) $2 \neq 4 \neq 5 \neq 6$ (5) $4 \neq 5 \neq 6 \neq 8$ (6) $4 \neq 5 \neq 7 \neq 8$

The line drawn between two elements of the mask shows a possible edge of a tetrahedron to be formed by this mask. While scanning the 3D Voronoi array, the mask looks for the boundaries of the four adjacent Voronoi polyhedrons. On detecting this situation, a tetrahedron is formed if one of the above six conditions is fulfilled.

Nevertheless, the above six conditions are not sufficient to prevent the erroneous formation in the case where more than four nodes are situated on the surface of a sphere. This situation is comparable to the 2D case shown in Figure 8.41 and Figure 8.43. Additional conditions are therefore added to prevent overlapping or intersecting tetrahedrons. Conditions (a) to (i) listed below correspond to the letters placed on the edges of tetrahedrons in Figure 8.44.

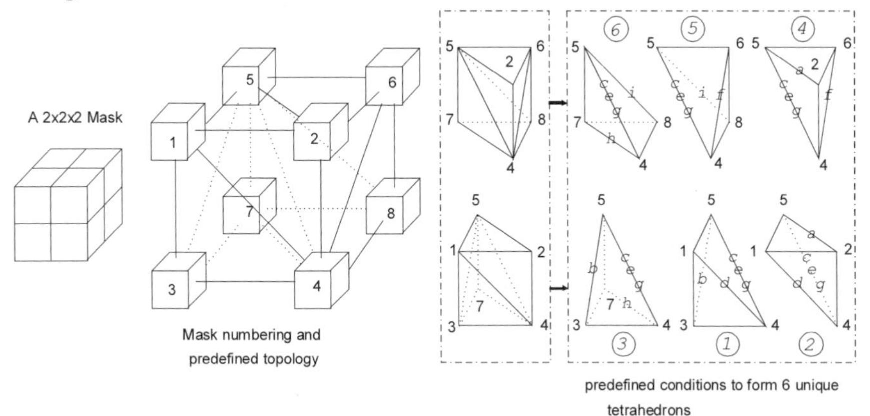

Fig 8.44 A 2x2x2 conditional mask for forming tetrahedral network.

(a) $1 \neq 6$, (b) $1 \neq 7$, (c) $1 \neq 8$,

(d) $2 \neq 3$, (e) $2 \neq 7$, (f) $2 \neq 8$,

(g) $3 \neq 6$, (h) $3 \neq 8$, (i) $6 \neq 7$.

Apart from the conditions above, another three must be added. These three conditions are needed to prevent the formation of a tetrahedron because of raster peculiarities. Similar to the broken appearance of an inclined line in a 2D raster image, in 3D an inclined plane appears as a staircase. This causes a problem when four points are situated on an inclined plane (relative to the 2x2x2 mask) and by chance on the circumference of a circle. Then, four adjacent Voronoi polyhedrons would be detected. Without the three conditions given below, a flat tetrahedron would be formed. This problem does not occur if there are completely horizontal or vertical planes (relative to the mask), since the previous conditions take care of such a constellation:

(j) not ($1 \neq 5$ and ($1 = 4$) and ($5 = 8$)),
(k) not ($3 \neq 4$ and ($3 = 5$) and ($4 = 6$)),
(l) not ($2 \neq 4$ and ($2 = 5$) and ($4 = 7$)).

Combining the first, the second and the third sets of conditions leads to the following algorithm:

if *(1) and (b) and (c) and (d) and (e) and (g) and (l)*
or *(2) and (a) and (c) and (d) and (e) and (g) and (k)*
or *(3) and (b) and (c) and (e) and (g) and (h) and (j)*
or *(4) and (a) and (c) and (e) and (f) and (g) and (j)*
or *(5) and (c) and (e) and (f) and (g) and (i) and (l)*
or *(6) and (c) and (e) and (g) and (h) and (i) and (k)*

then

increase number of tetrahedrons (for memory allocation)
form a tetrahedron.

Since there are several alternatives in designing this mask (for example, a cube can be decomposed into five or six tetrahedrons), the mask in Figure 8.44 takes into account the compatibility with 2D triangulation. This provides an easy way of combining a TIN with TEN without any conflict. The design of the 2x2x2 mask is based on the principle that a cube can be cut by three different planes, each plane passing through two diagonally opposite edges of the cube, taking a pair of edges for every coordinate axis. The three planes intersect each other along a diagonal of the cube and divide the cube into six tetrahedrons.

The topology of the resulting simplicial network is documented by tables as shown in chapter 6.

Composition of Features

To complete the construction of the 3D spatial model, the topological relationships between the simplices and the complexes must be established by classifying (assigning) each simplex as part of the complex it constitutes. This can be achieved by performing an overlay process between the data sets containing features and the simplicial network respectively. As a result of the convex property of simplices, the centroid of each simplex can be used for the containment test against the complex. This significantly simplifies the overlaying process. Note that the overlaying process requires both data sets to be correctly structured in advance. The most favourable data structure for the features is of that derived from the variants of FDS (see chapter 4), because of the compatibility.

Data Structuring for 3D FDS

The construction of a 3D spatial model based on a simplicial network requires the incorporation of 3D features as constraints. The constraints are preferably structured in the database according to 3D FDS. This database provides the geometric components of the constraints involved in the network construction and the thematic components for the composition of features by the overlaying process. A relational data structure derived from 3D FDS has been presented by Rikkers et al., (1993), Bric (1993) and Bric et al., (1994).

In Bric (1993), the capability of 3D FDS has been explored by building TREVIS an experimental 3D GIS. Various queries about topological relationships between features in 3D space, for example neighbourhood, adjacency and inclusion, have been tested with satisfactory results. The data sets used for the experiment were, however, structured manually. The complexity of constructing a database of 3D features was realized and simplification of the process by capturing necessary spatial relationships at the data acquisition phase was proposed. Since aerial photographs were to be used as the data source, strict photogrammetric digitizing procedures were suggested. Further investigation into the design of photogrammetric digitizing procedures has been carried out by Wang (1994), placing the main focus on urban scale application. Database creation involves the reconstruction of 3D features representing buildings, houses, and surface features, for example roads, land parcels, and terrain relief. Each digitized feature was assigned a specific code indicating for which data the structuring strategy was suitable, as shown in Table 8.1.

Table 8.1 Examples of output format from photogrammetric data collection process

Type	Code	Description/Example	Purpose
Body feature	B1	Roof outline	Construct the body by plane sweep vertically to intersect with DTM
Surface feature	S1	Roof facet boundaries (ridge and drainage)	Replace the roof outline after obtaining the body
Surface feature	S2	Land parcel	Make part of terrain surface
Line feature	L1	Road, railway	Make part of terrain surface
Point feature	P1	Location of a tree, lamppost	Make part of terrain surface

Note that only a limited number of codes were made available. Two major groups of codes can be distinguished, the first contributing to the representation of buildings and the second contributing to the representation of terrain surfaces. Since the aerial photographs are limited to near vertical view, roofs of buildings and houses, 3D objects of interest can be captured quite easily. The knowledge and experience of the operator carrying out the digitizing are important for the correct interpretation of the situation shown in the aerial photograph.

The data structuring can now proceed as shown in Figure 8.45. All possible topological relationships are also recorded in parallel with the reconstruction of the geometry using the data structure (see Rikkers et al., 1993)..

Representation of the terrain surface and outline of the roofs are used to construct the 3D features representing the buildings and houses. The footprints obtained are incorporated in the 2.5D simplicial network, that is, a TIN which represents the terrain surface and 2D features shown in Figure

8.45 (e) and (g), respectively. In this way, the topological relationships between terrain surface and 3D features can be established and maintained within the 3D FDS database. In addition to the construction of geometry in Wang (1994), the process shown in Figure 8.45 also includes face orientation, fulfilling the 3D visualization requirement for the normal vector of each visible face of a body feature to point towards the outside of the body.

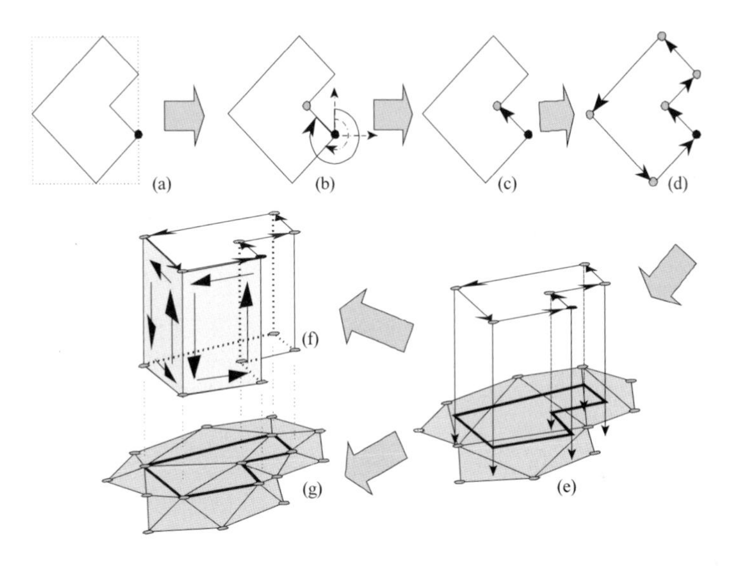

Fig. 8.45 Steps for data structuring for 3D FDS.

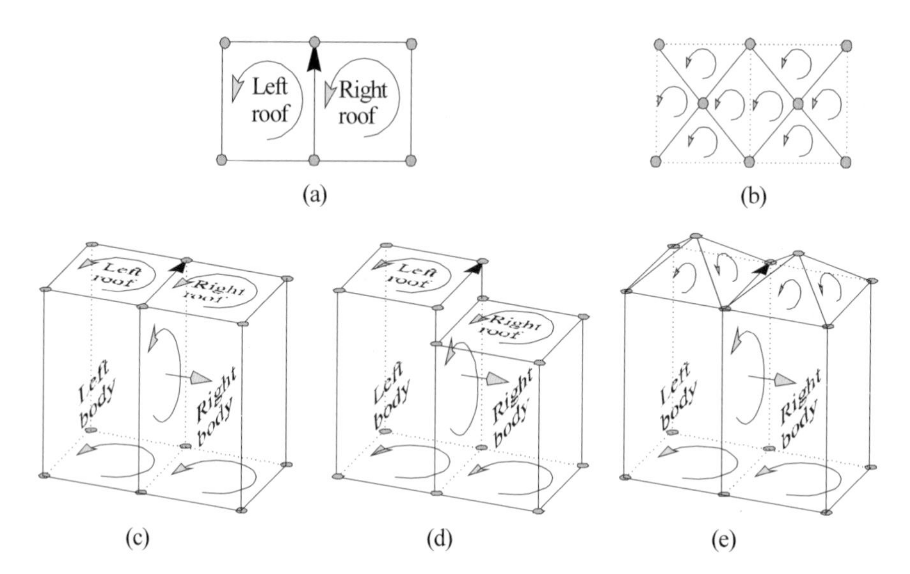

Fig. 8.46 (a) Outline of two adjacent roofs, (b) details of the roof, (c) two adjacent flat roofs with the same elevation, (d) similar to (c) but different elevation, (e) replacing the outline of the roofs by their facets after reconstructing the main geometry

Note however that not all topological relationships as described in Molenaar (1990) (for example a node in a body) can be established by this data structuring through the lack of information during data collection. Some adjacency may be capable of being directly inherited from 2D topology (see Figure 8.46 (a) and (c)). Nevertheless, the relationships may not always be straightforward, for example as shown in Figure 8.46 (d). This implies that further investigation of this issue is still needed.

8.11 Summary

In this chapter, several important algorithms for TINs (2D and 3D) constructions have been introduced, namely the DT, the Voronoi tessellations, triangulations (including constrained triangulations), and constrained line rasterisation. One of the GIS application algorithms has also been introduced, that is the contouring. Other major task in the development is TIN data structuring, and topological data structuring for 2D TIN and 3D TIN (TEN) are also described.

We have presented methods for data structuring, database creation and internal consistency checking for the building of an integrated 3D database based on IDM and UNS. We have also outlined the procedure for acquiring 3D urban data and structuring them according to 3D FDS, an intermediate step in the construction of an integrated 3D spatial model. For generating a simplicial network with constraints, a vector approach is very cumbersome. We have therefore proposed and developed a raster approach which simplifies the implementation. Note, however, that this algorithm has not been designed to adapt locally to different densities of nodes. Special care should be taken during the rasterization process where the selection of an appropriate raster cell-size is suggested, to avoid information loss. Although raster processing is simple and fast, it requires a large amount of memory and storage space. This problem is becoming less significant with the rapid development of both computing power and storage capacity.

Since the raster approach forms a simplicial network from Voronoi regions, we have developed an approach that is valid for nD, incorporating constraints into the simplicial network using invariant property of Voronoi regions under Voronoi tessellation. This completes the geo-spatial modelling ranging from the design to the construction phase.

Creating an integrated 3D database requires many steps. When contemplating a large area with highly detailed information, the task may seem impossible. We therefore suggest incremental construction of the integrated database. This is one of the most important aspects that the simplicial network structure offers. Also, more detailed information can be contained through network refinement; this can be done locally. Although the proposed data structuring is attained by raster processing, the end result is a vector structure that does not depend on scale, or level of precision. The simplicial network can be refined as necessary, so it can be used to model broad ranges of real world objects.

To certify the integrity of a generated simplicial network, the generalized Euler equality derived in chapter 6 can be used for the internal consistency checking, valid for the geometric aspect of the simplicial network. Further checks may still be necessary. However, more in-depth investigation is needed to cover this aspect for the maintenance of the simplicial network.

Chapter 9 APPLICATIONS OF THE MODEL

Having designed the integrated data model, unified data structures and introduced methods of construction, demonstrating the 3D spatial model's applicability is the last objective. This is achieved through various steps of spatial data processing, using both simulated and real data. It is a stepwise approach, starting from the 2.5D application that integrates terrain relief and terrain features. Examples of spatial query and 3D visualization exploiting an integrated spatial model based on a simplicial network are also presented. For the full 3D application, two kinds of data sets are used as examples. The first uses the data of an urban area consisting of roads and buildings. The second uses a simulated boreholes data. The latter is used to demonstrate the modelling of subsurface objects typically found in geological applications.

9.1 Integration of Terrain Relief and Terrain Features

The combined use of terrain relief and terrain features has been often referred to as 'integration of DTM and GIS' as mentioned in chapter 1. This section intends to demonstrate that the concept of a simplicial network can offer a solution to this problem. The general aim is to create an integrated spatial model representing the earth's surface and 2D representation of spatial objects related to this surface in a 3D space. Simulated data sets are used for this demonstration. The data sets consist of:

- Surface features shown as land use and soil maps in Figure 9.1 and Figure 9.2, respectively
- Line features shown in Figure 9.1 and Figure 9.2
- Measurements for terrain relief modelling shown in Figure 9.3.

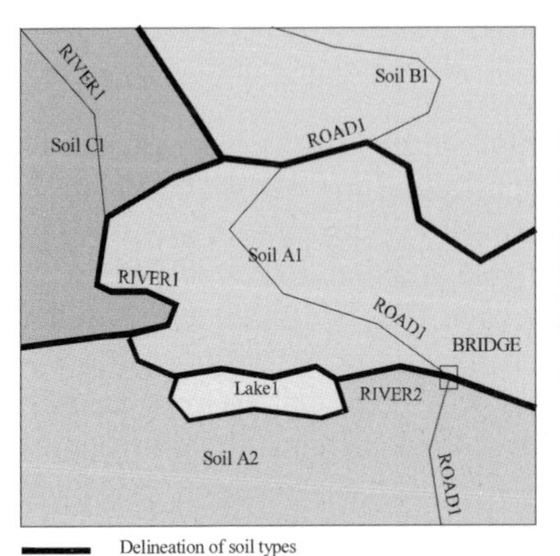

Fig. 9.1 land use map.

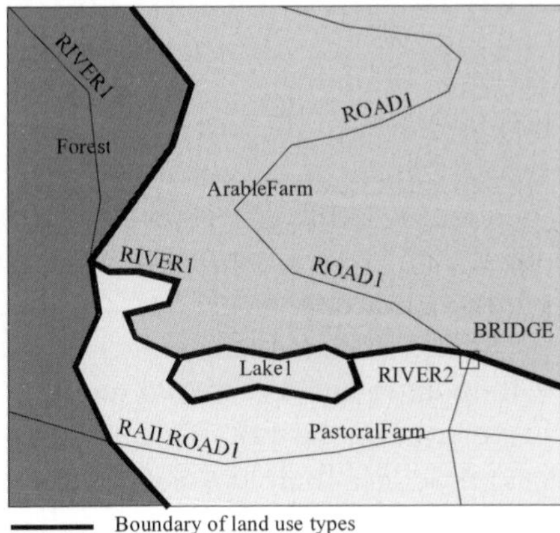

Fig. 9.2 Soil map.

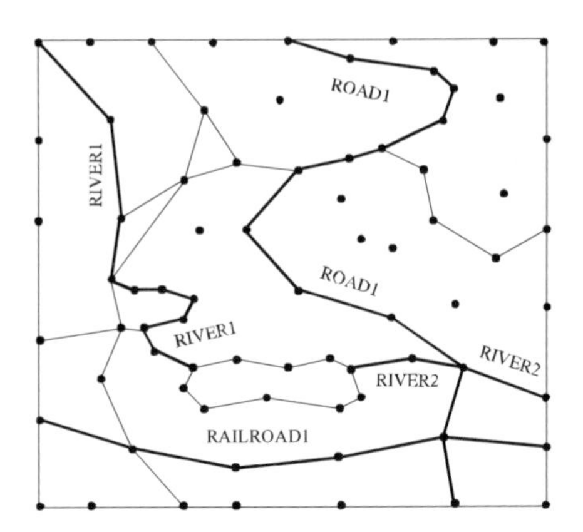

Fig. 9.3 DTM points. Line features and boundaries of area features to be used as constraints.

The data sets of land use, soil, line features and relief may be used to answer the questions listed below:

- Which sections of the Road1 may need side slope protection?
- Where is a suitable location for dam construction?
- What is the total length of the railroad that may be damaged by a 1-metre flooding from Lake1 this year?
- How large is the surface area of the soil type A1 between elevation 100 and 200 metres?
- Where are the forest areas with a terrain slope more than 30%?

- How large is the catchment area generating run-off into the River1?

9.2 Creating an Integrated Database

The maps shown in Figure 9.1 and Figure 9.2 are in the form of SVVM. They should be combined by overlaying in the first step of the integration. This yields a multi-theme data set shown as a multi-valued vector map (MVVM) in Figure 9.4. The multi-theme data set is used as a basis to create relationships between features and primitives at a later stage. Since there are common features in both land use and soil data sets, that is, roads, rivers, railroads and lakes, redundancy and uncertainty problems arise. For example, the digitizing of Road1 in land use and soil data sets may be carried out separately. Overlaying these two data sets may introduce slivers (small polygons) along Road1. This problem, as shown in Figure 9.5, should be solved before proceeding further.

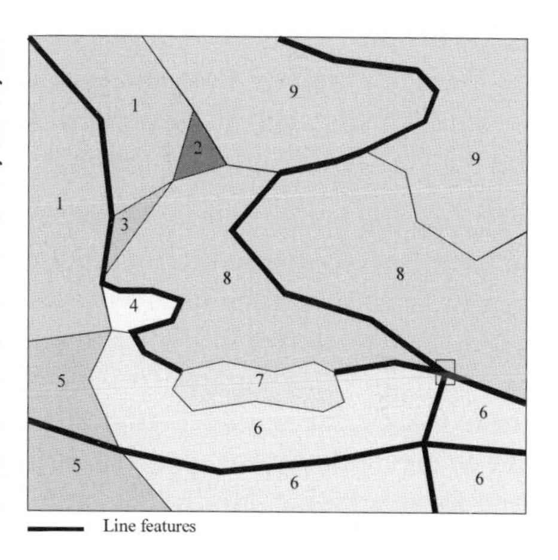

Fig. 9.4 Overlaying of land use and oil map result in a multi-theme map

Shi (1994) has discussed the handling of this kind of uncertainty. It should be noted that uncertainty for both planimetry and height are likely if all data points have 3D coordinates. However, it has been assumed here that this problem has been solved, allowing us to proceed further.

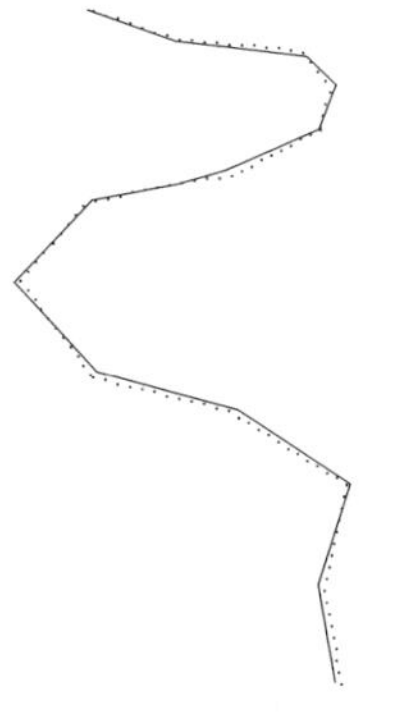

Fig. 9.5 Slivers caused by uncertainty in data acquisition may occur after overlaying of different data sets. A solid line and dotted-line show two different sources of data for the same line feature

The next step is to ensure that all digitized points have 3D coordinates. If some of these points have only 2D coordinates, a preliminary Delaunay triangulation may be constructed using only the nodes that have 3D coordinates. Linear features that are already in 3D should be involved as constraints in the triangulation process, to ensure better fidelity of the surface representation. The TIN resulting from the constrained triangulation in this step can be used to introduce the third dimension to all 2D data points by means of interpolation. Each 2D point can be tested with the point-in-triangle algorithm. The plane equation can be computed from the triangle obtained from the positive results of this test. The height information is obtained at the point of intersection between the triangle plane and the vertical line passing through this point.

When all the nodes with 3D coordinates have been obtained, they should be used for constrained triangulation. All line features and boundaries of area features should be used as constraints, as explained in chapter 8. The resulting TIN, shown in Figure 9.6, is then overlayed with the multi-theme data set, shown in Figure 9.4. The relationships between triangle components, that is, faces, edges, nodes, and the point, line and surface features can be created after this process. It is worth mentioning that the algorithm used for overlaying TIN with a polygon map can be simplified to point-in-polygon testing. The centroid of each triangle can be used as a point to be tested against a set of polygons.

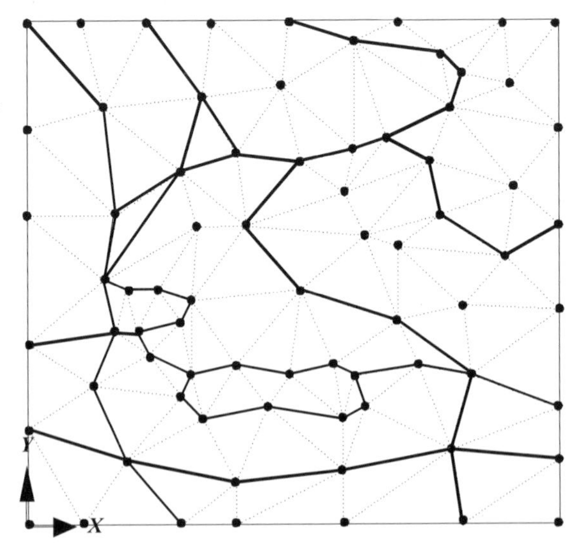

Fig. 9.6 TIN resulted from constrained triangulation using all digitized points. Line features and boundaries of area features are used as constraints

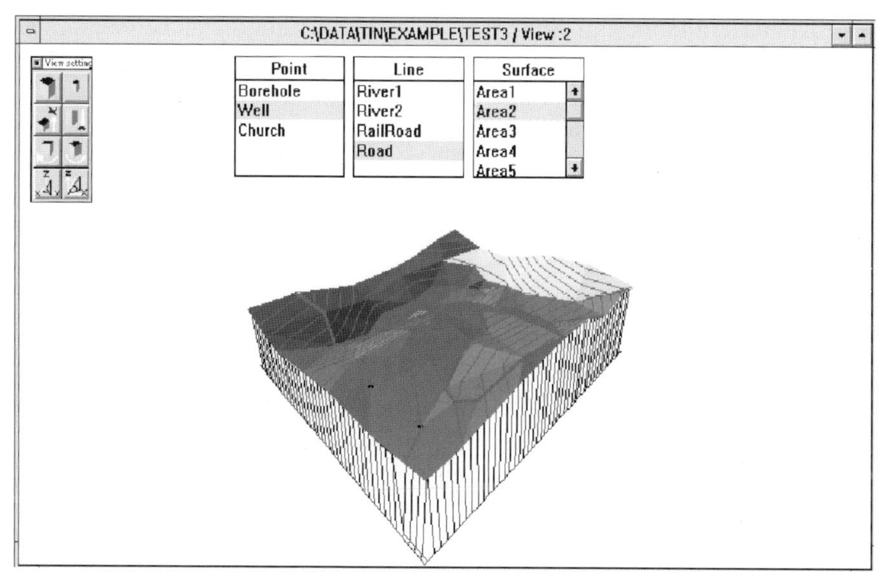

Fig. 9.7 The constrained TIN presented in a perspective view shows aspect of relief of this data set. The query results can be directly presented in this view. Contour lines can also be derived directly from this database

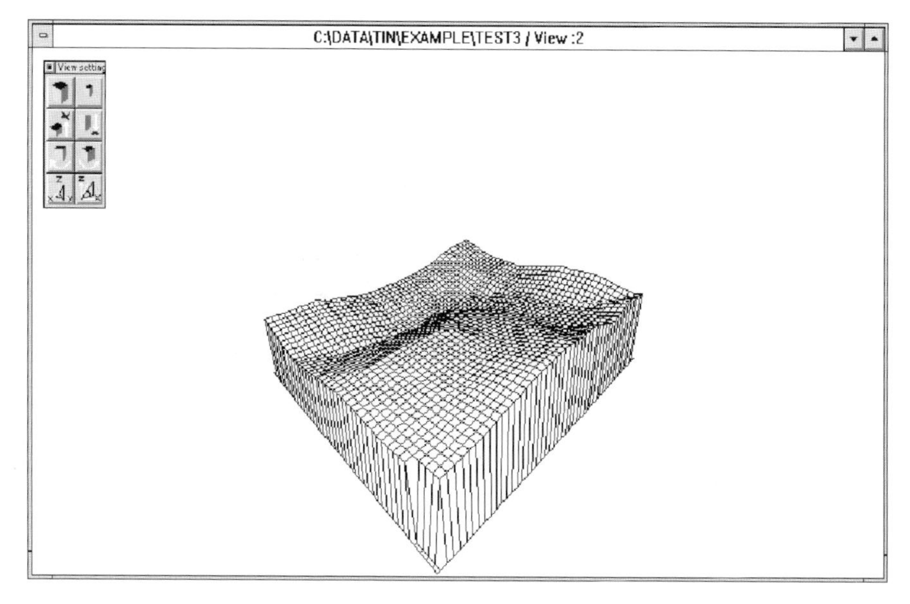

Fig. 9.8 Regular-grid DTM derived from simplicial network database

The database obtained at this stage is called a 'simplicial network integrated database' (SNIDB) for convenience. Figure 9.7 presents the content of the SNIDB in perspective view that creates understanding about relief. Data structured in this way permit direct presentation of query results in this kind of view. Regular-grid DTM can also be derived from this database, as shown in Figure 9.8.

9.3 A Spatial Query Example

Having the database in a SNIDB scheme extends the query space typically provided by a 2D GIS or a DTM significantly. Many complex queries requiring many steps when using a typical 2D GIS can now be simplified. For example, a road engineer looking for soil material of type A1 to use for road construction might ask, 'What is the volume of soil A1 within the arable farm area and at a depth of 3 metres under the average elevation of this area?' Using the SNIDB, the system just looks for all the triangles that are part of the polygons having feature code = 8, computes the mean elevation using this set of triangle vertices and then computes the summation of volume under each triangle with a depth of 3 metres below this mean elevation. The volume under each triangle can be computed using the formula shown in Figure 9.9. The process is illustrated in Figure 9.10.

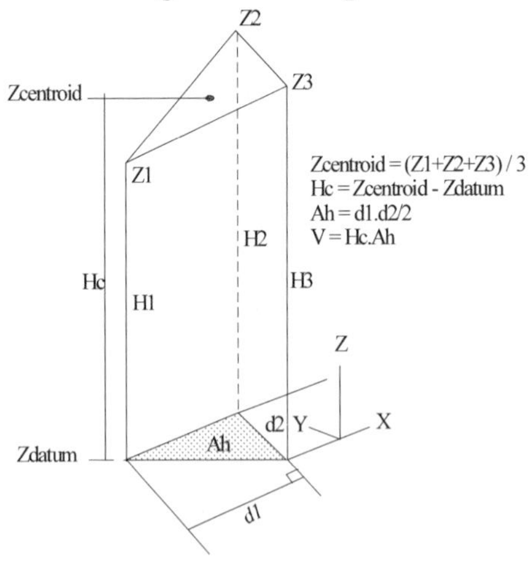

Fig. 9.9 Computation of volume above datum and under a triangle.

Without the SNIDB, a typical user of several databases:

(a) overlays the soil database and the land use database to obtain the overlapping area of soil A1 and arable farm area.
(b) solves uncertainty, for example in the form of slivers.
(c) overlays the results obtained from (b) with the DTM database to obtain the clip DTM within the area of soil A1 and arable farm area.
(d) uses the clipped DTM, calculates the mean elevation within the area.
(e) calculates the volume.

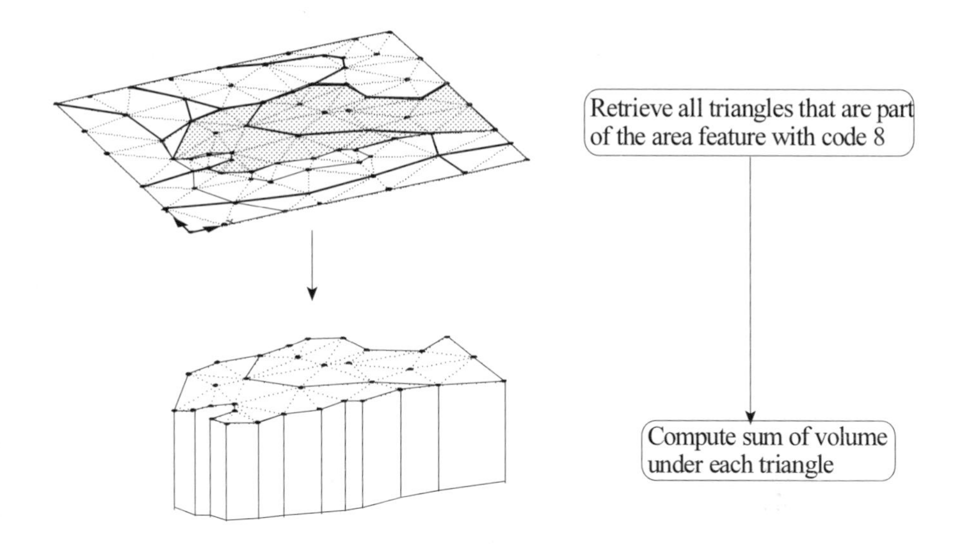

Fig. 9.10 Computing summation of volume under and area tessellated into a set of triangles.

The above requires two subsystems, that is, 2D GIS and DTM. Steps (a) to (b) are carried out in 2D GIS whereas steps (c) to (e) have to be done in the DTM subsystem. Two overlaying processes are required, entailing extra time to process the query.

The integration into SNIDB takes the responsibility of overlaying proccesses away from the user. Although constructing the SNIDB might take more time, the user gains the response time during the query process, which seems to be more reasonable, because the query tends to take place more frequently than the database construction. For example, the same kind of question may be asked again for a different area, elevation and date, requiring a repeat of the process from (a) to (e). This repetition is needed for every different set of parameters given.

The overlaying process normally requires computational geometry at the level of geometric primitives. A large amount of operating time may be required, especially for a large data set. The second problem is the solving of uncertainty, which requires knowledge about the source of data and the appropriate solution to be taken. The SNIDB approach performs preoverlaying at the database construction level and therefore turns the overlaying process during a query into a spatial search which can be speeded up by the use of topology. The uncertainty only needs to be solved once and is then converted into data quality that can be stored as an attribute in the database. These features make the GIS more convenient to use.

9.4 Integrating with 3D Features

Should the model need to cover the full three-dimensional representation of buildings and other man-made objects as shown in Figure 9.11, the simplicial network can also meet this requirement. The assumption is that the representation of 3D objects is given in the form of 3D FDS. Since the 2D simplicial network is fully compatible with the 3D FDS, the complexity of integration is reduced. As a result of compatibility, a feature belonging to the 2D simplicial network can readily be considered as a feature of 3D FDS. A surface feature (that is, part of the terrain surface) can be related with a 3D feature via their footprints. This can be done by embedding the footprints of 3D features within the simplicial network representing terrain surface by means of constrained triangulation, as shown in the chapter 6. The footprints of the 3D features carry the links to the terrain surface and the 3D features themselves. In this way, the 3D topology between the surface and 3D features is established, which permits the integrated use of the two types of data within a 3D FDS database. Figure 9.12 shows an example of terrain data. Figure 9.13 shows the data in the form of a simplicial network that has been constructed by constrained triangulation.

Fig. 9.11 Scanned aerial photograph of the study area of the city centre of Enschede, The Netherlands.

Fig. 9.12 2D features on the terrain surface of the study area. The data was photogrammetrically digitized using a Matra T10 Digital Photogrammetric Workstation.

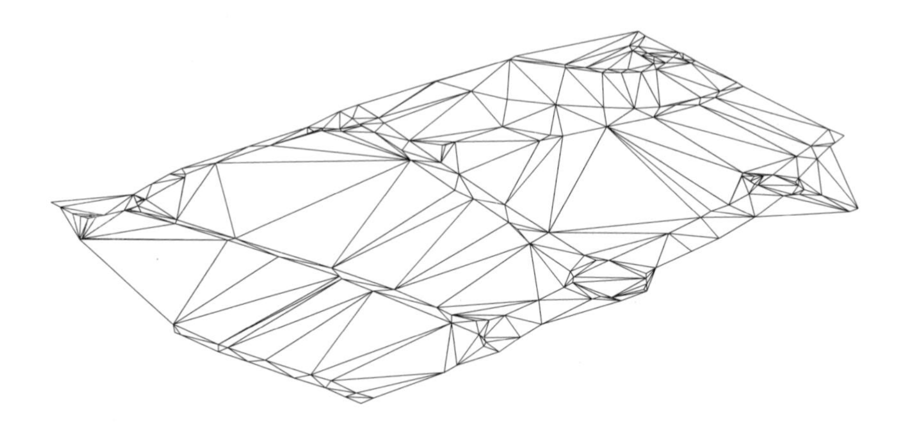

Fig. 9.13 The result of triangulation applying 2D features as constraints. The constrained triangulation was carried out using the raster approach implemented in the ISNAP program.

Figure 9.14 shows the data of 3D objects. The data has been digitized by the stereo photogrammetric approach using a Matra T10 digital photogrammetric work station. Only the outlines of the roof of each building were digitized manually.

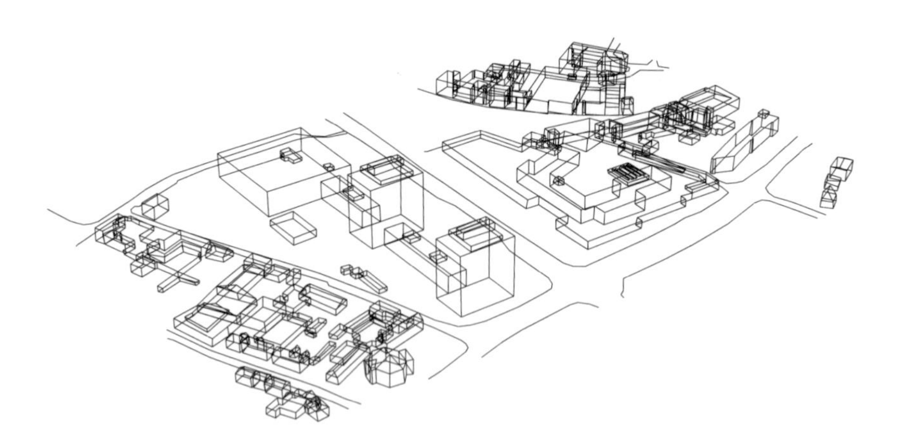

Fig. 9.14 Buildings with walls and footprints resulting from vertically projecting the roof outlines onto the TIN-DTM.

The wall and footprint of each building were obtained automatically by vertically projecting the outline of the roof onto the TIN-DTM, shown in

Figure 9.13, using a set of programs developed by the author and Wang (1994). The 3D objects are maintained using the 3D FDS scheme.

The footprint of each building is then retrieved from the 3D FDS database, as shown in Figure 9.15.

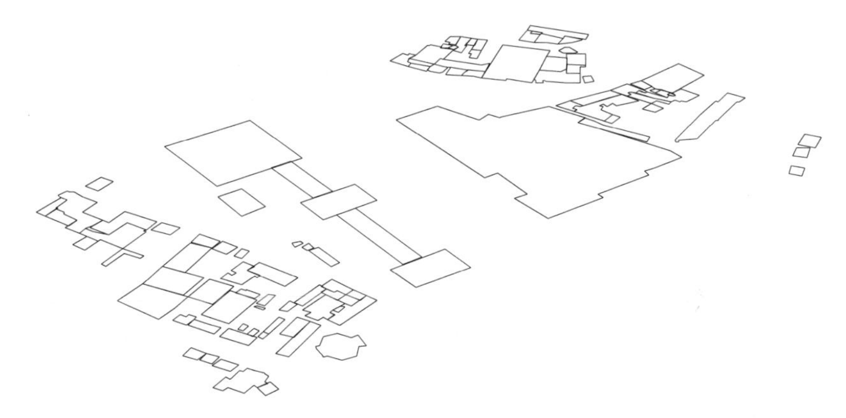

Fig. 9.15 All footprints of the buildings extracted from the 3D FDS database.

By mean of constrained triangulation or local updating of a 2.5D simplicial network, all footprints of the buildings are embedded onto the terrain surface that is represented by a simplicial network, as shown in Figure 9.16. By means of overlaying, the part-of relationships can be established between triangles and a surface feature representing the footprint of a building and embracing them. All triangles affected by retriangulation are also subjected to updating the part-of relationships with the corresponding surface features.

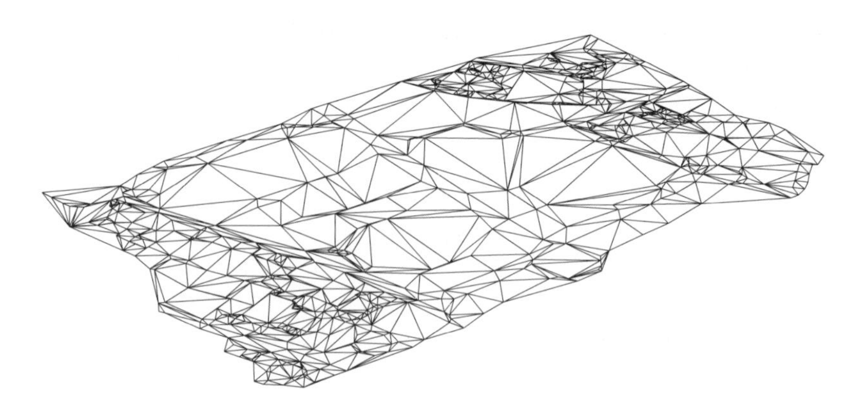

Fig. 9.16 A 2D simplicial network representing a terrain surface of part of the central area of Enschede, The Netherlands.

Figure 9.17 shows the final result of the integration between the surface and 3D objects.

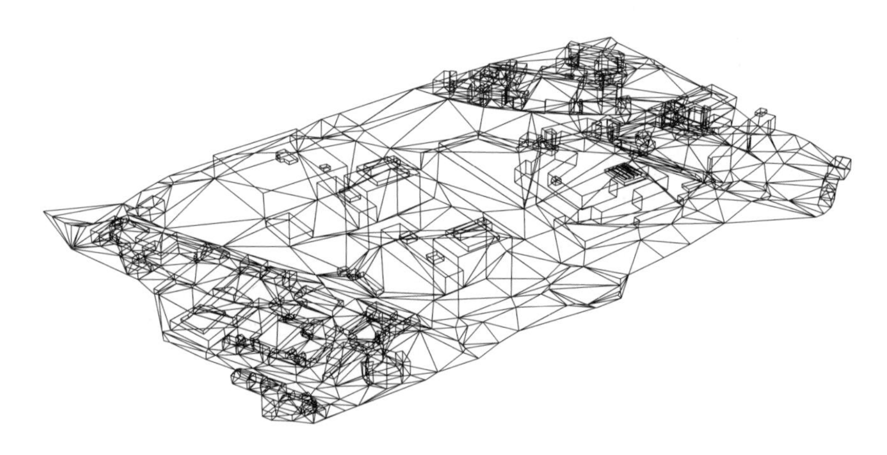

Fig. 9.17 Merging of the representations of terrain surface and the 3D objects.

This kind of database may be used to answer questions such as:

- Which buildings are suitable for the placing of antennas for mobile telephones?
- Which buildings are visible from point A located on the top of building B?
- Will the noise from the high-speed trains passing nearby be heard in this city?

If the representation of a terrain surface, in which the footprints of 3D features may be included, is stored separately from the database storing 3D features, the user may have to face several problems before a query can be carried out, for example:
- transformation to a common coordinate system
- solving uncertainty with the footprint of a building in the 3D database if the footprint is also on the 2.5D database.

9.5 Integrating with Geo-scientific Data

Geo-sciences and engineering, including geology, air, soil and water pollution control, civil and geotechnical engineering, require the investigation of objects of interest at levels beyond the scope of data sampling. Mathematical models involving finite element analysis are typically used. The SNIDB can facilitate such a requirement, because each geometric primitive has a finiteness property. A spatial model in the form of a 3D simplicial network may be used to answer questions such as:

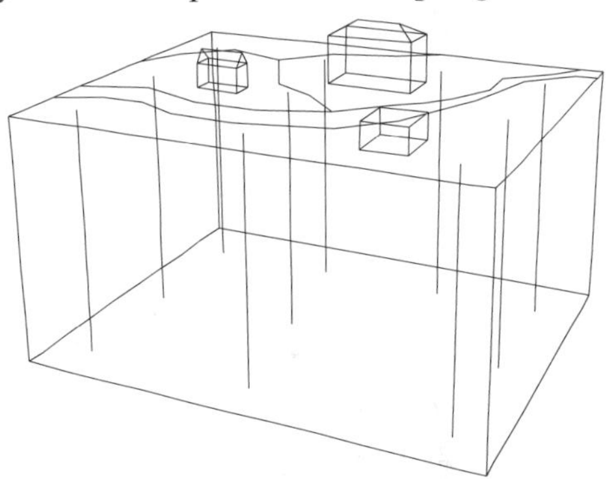

Fig. 9.18 An example of borehole locations for sampling geological data. Vertical lines indicate boreholes.

- Does the clay layer extend over the entire area under the construction site?
- Which buildings are potential sources of chemical disposal into the ground?
- Will these buildings be affected by the excavation of the soil to 10 metres depth from the ground surface?

Data obtained from boreholes, as shown in Figure 9.18, usually have at least one measurement in addition to the coordinates (x, y, z). The data can be stored within the SNIDB, as it is for further processing in the spatial analysis.

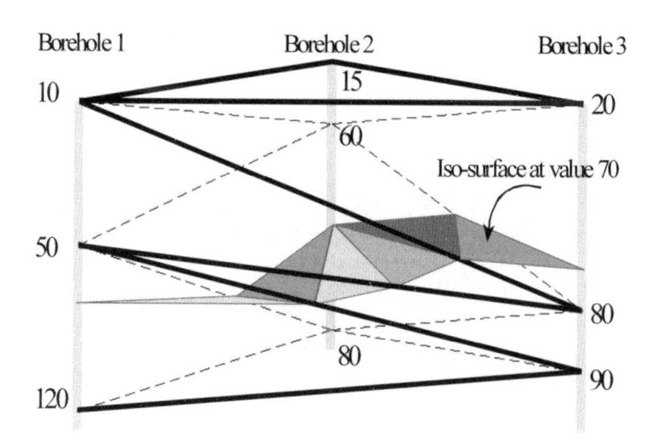

Fig. 9.19 Derivation of iso-surface from a tetrahedral network constructed using three boreholes. The numbers in this figure are property values used for interpolation of the surface.

In this case, the SNIDB must apply the tetrahedral network to structure such data. All point, line, and surface objects encountered at this stage must be involved in the tetrahedronization as constraints. The SNIDB can support a data base query for those objects as well as for some finite element analyses such as volume, bearing capacity, soil strata, and the like.

Figure 9.19 shows how an iso-surface can be derived from a TEN. The surface can be obtained in the form of a TIN.

Figure 9.20 is a wireframe plotting of the 3D simplicial network generated from the simulated borehole data, using the ISNAP program (see section 7.2.6). Figure 9.21 shows the derivation of upper and lower surfaces bounding a soil stratum from the generated 3D simplicial network.

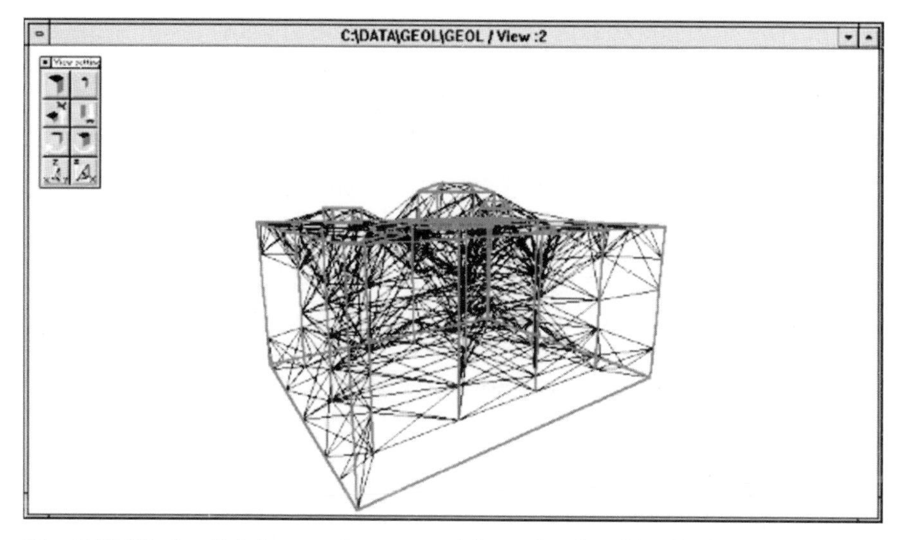

Fig. 9.20 3D simplicial network generated from simulated borehole data

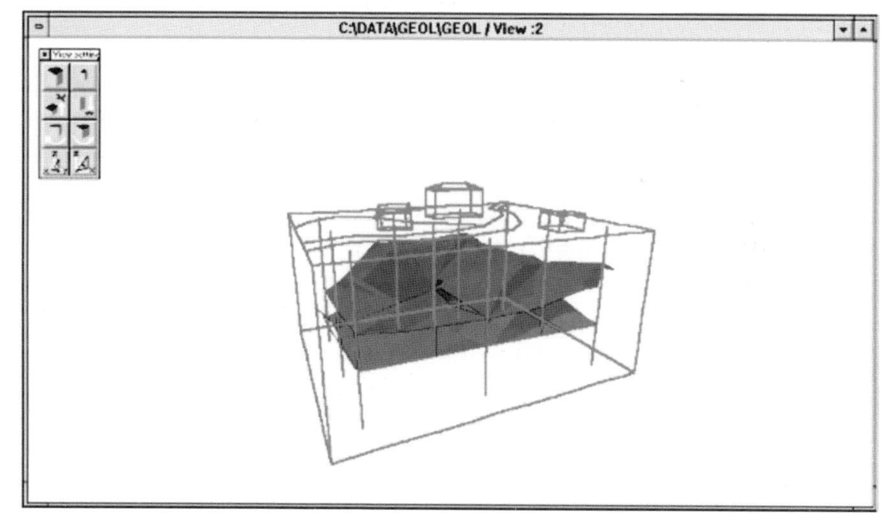

Fig. 9.21 Derivation of iso-surfaces from a 3D simplicial network

9.6 Spatial Operators

The applicability of the designed spatial data model also depends on the user interface (for example, the query language) and the availability of spatial operators. Both of these applications define the functionality of the system. The spatial operators define the operation at a low level, while the spatial query language defines the operation at a relatively high level capable of being well understood by human beings.

The spatial operators are comparable to those found in mathematics, set and logic algebra. The basic operators are union, intersection, or, and, xor, and so forth. These basic operators can be combined to build more sophisticated functions. Different kinds of spatial relations, that is, metric, order and topology, are also essential to the design of spatial operators. Metric operators are built around the computational geometry, for example point-in-polygon, point-in-body, intersection of lines, and intersection of surfaces. Metric operators can be used to derive topological relationships. With respect to the simplicial network data model, these problems can be reduced to the level of a simplex. Point-in-polygon and point-in-body can be reduced to point-in-triangle and point-in-tetrahedron respectively. The intersection of bodies can be simplified to intersections between tetrahedrons, which can be further reduced to intersections between simplices of a lower dimension, for example between triangles, edges, as shown in Figure 9.22.

Order operators are those used to compare and arrange spatial elements. Topological operators are those defined by topological relationships, like containment, touch, coincidence, disjoint, left, right. For example, a body feature A is contained in another body feature B if all tetrahedrons of A are contained in B. Body A is a neighbour of Body B if a tetrahedron of A is a neighbour of a tetrahedron of B. Bodies A and B are coincident if all tetrahedrons of A are tetrahedrons of B, and vice versa. The spatial operator is essential to spatial analysis. It is a link between the spatial query language and the spatial analysis function.

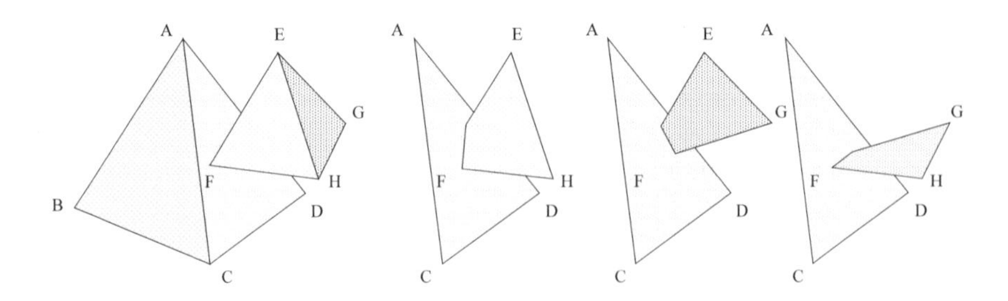

Fig. 9.22 The intersection of two tetrahedrons can be reduced to intersections between triangles or intersections between edges and triangles.

The concept of a simplicial network also helps to simplify the development of 3D computational geometry. All spatial elements in the data model have finite properties. The triangle and tetrahedron are convex geometric shapes, making many complex computations simpler. Many 2D operations can be readily generalized into 3D. For example, the algorithm for point-in-triangle can be generalized into point-in-tetrahedron. An algorithm to compute the area of a triangle can also be generalized into computing the volume of a tetrahedron. Generalization into n-dimensions is implied.

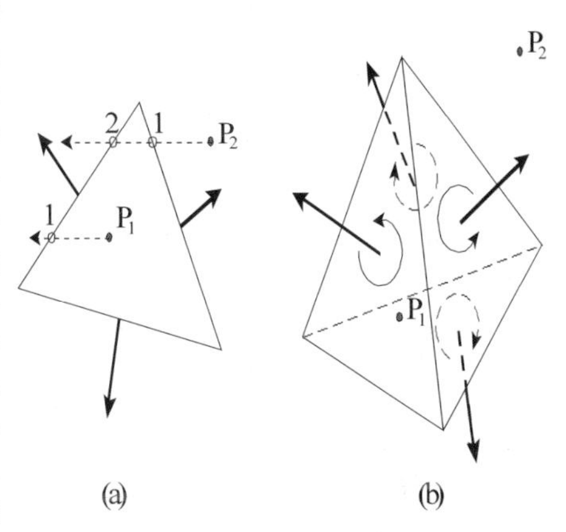

Fig. 9.23 Point-in-triangle and point-in-tetrahedron testing.

Figure 9.23(a) shows an algorithm for point-in-triangle testing. A point P1 situated inside a triangle is always in the negative direction of the normal vector of each edge of the triangle (given that the normal of each edge points outwards from the triangle). This is not the case for a point situated outside the triangle. This assumption also holds for testing if a point is contained in a tetrahedron as shown is Figure 9.23(b). Note that this algorithm is not valid for a non convex polygon or polyhedron. The line-intersection test can be used by counting the odd or even numbers of intersections of the line emanating from the point (see the dashed lines in Figure 9.23(a)) with the boundary of polygon or polyhedron. An odd number indicates that the point is inside the polygon or polyhedron, while an even number indicates that the point is outside.

Spatial operators to calculate some properties of the spatial object, for example volume, surface area can be designed more easily. The volume of a complex object is the summation of the volumes of all the tetrahedrons that are part of the object. The surface area of the complex object can be computed from the summation of the area of all the triangles that are part of the boundary of the object.

9.7 Graphic Visualization

In 3D geoinformation, visualization is one of the most important components of the system. Realism and interaction are necessary for information to be quickly understood. The key is speed of data processing, which relies on the power of the system and an appropriate data structure. SNIDB permits the visualization of the representations of both determinate and indeterminate spatial objects. The representation of determinate spatial objects can be displayed directly, while the derivation of the boundaries is needed for indeterminate spatial objects prior to their graphic visualization. A simplicial network supports different types of graphic visualizations, as described below.

9.7.1 Wireframe Graphics

Wireframe graphics give a relatively low level of realism. They only make use of nodes and arcs stored in SNIDB. Without interactivity, wireframe graphics seem not to be very useful for complex, or large amounts of data (refer to Figure 9.19, Figure 9.20). The operation to display 3D wireframe graphics consists of transforming all the coordinates of the nodes into a perspective system relating to the observer and the viewing distance to the objects.

With all the coordinates in the perspective system, the next step is to use arc and node topology to draw the straight lines connecting the beginning and end nodes of each arc. The wireframe graphic is then obtained. Examples of wireframe graphics are shown in many of the figures in this chapter. When there is a need to differentiate different type of graphically displayed information, different styles, colours and line thickness can be used.

Visualization using wireframe graphics can be further improved by adding stereoscopic vision capability. A simple and economic approach is anaglyphic stereo, using red and blue (or green) filters to separate two parallax images from the viewer's left and right eyes. The parallax images are displayed using different camera positions along the line parallel to the eyebase of the viewer. The blue (or green) shade is used for the left image and the red shade for the right image. The viewing glasses must have red and blue (or green) colours in the opposite sense of the displayed images. The perception of depth helps to resolve visual ambiguity on a 2D display. Figure 9.24 is an example of an anaglyphic stereo pair.

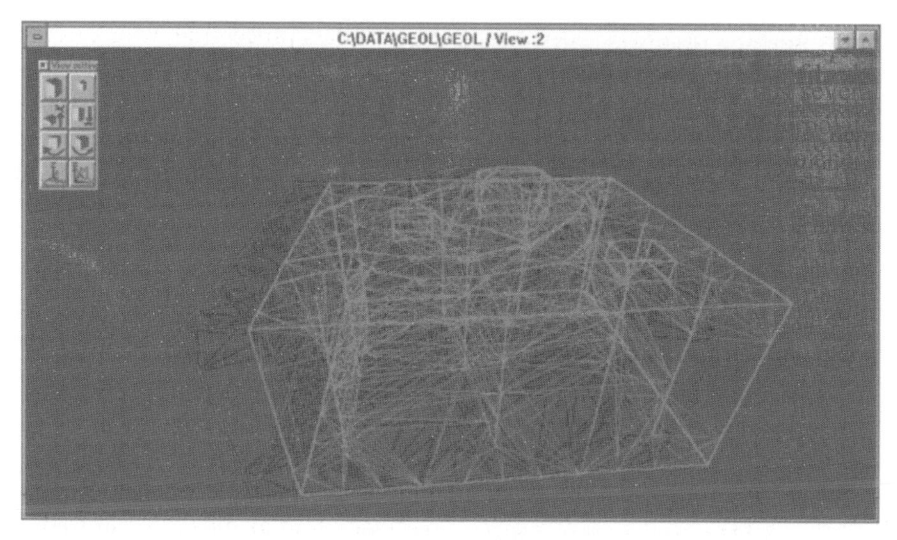

Fig. 9.24 An example of wireframe images displayed in anaglyphic stereo mode.

9.7.2 Hidden Line and Surface Removal

The 3D visualization with wireframe graphics can be significantly improved by applying a hidden line and surface removal operation. For this purpose, many algorithms are available in computer graphics and CAD (Beaty and Booth ,1982; Foley et al., 1992). One of the most efficient and relatively simple algorithms is known as a 'z-buffer' (or 'depth-buffer') which is the raster-based operation that only stores the pixels belonging to the visible part of the objects in the scene. For each location on the buffer that is a 2D array, only the pixel nearest to the viewer is stored. Each pixel value indicates the identifier of the facet. A simplicial network provides information in the form of a triangle that can be used for the purpose. However, some extra spatial index structure (for example, a BSP-tree) needs to be constructed on top of the core database to speed up and ease the operation. For example, if the depth sorting algorithm is used, all triangles need to be sorted according to the distance from the viewer. The remote triangles are displayed first, while the triangle closest to the viewer is the last to be displayed. Each triangle must be filled by background colour, while its boundary is drawn in foreground colour. In this way, part of the objects that should be invisible are overwritten. Only the visible parts remain on the display device. The z-buffer is also a kind of index structure in the form of a regular grid.

9.7.3 Surface Shading and Illumination

When the hidden line and surface removal operation described above has been applied, surface shading and illumination can take place next in the sequence. Colour can be assigned to each triangle and then displayed by a filling operation during the hidden line and surface operation. For surface illumination, the lighting model must be used to compute the colour intensity for each triangle. The intensity depends on the amount of light reflecting from the facet to the viewer. The lighting models available are Gaurad shading or Phong illumination (Foley et al., 1992). The general light geometry is shown in Figure 9.25.

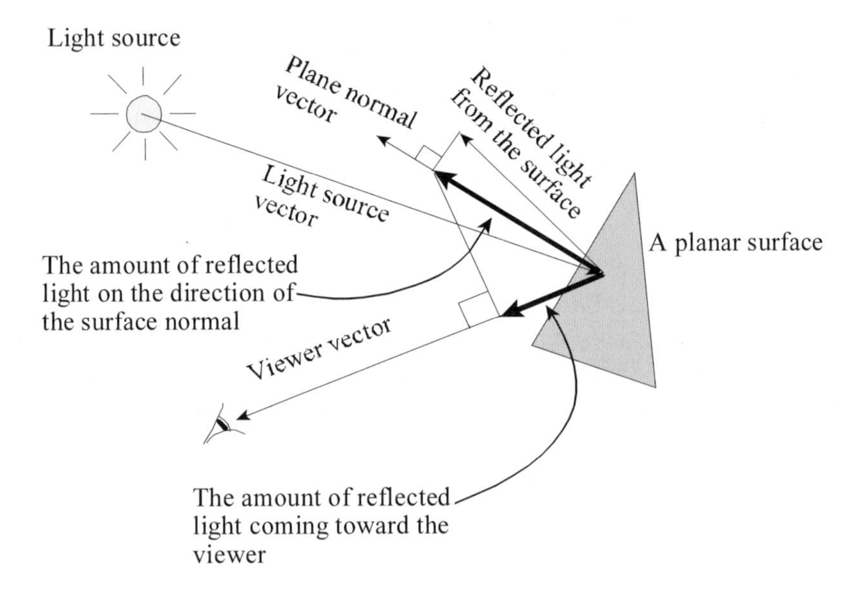

Fig. 9.25 Principle of surface illumination

A simplicial network provides the planar surface that is ready to be used for the calculation of a plane's normal vector. Since the order of vertices of the facet determines the direction of the normal vector, it is advisable to store this information systematically in the database. For example, all triangles belonging to the earth's surface should have their normal vectors pointing upwards so as to be able to interact naturally with simulated sunlight. To calculate the amount of light reflecting from a triangle to the viewer, the reflected light from the surface is first projected onto the normal vector of the triangle. This projected light is then projected onto the viewer vector which determines the intensity of light the viewer perceives from this light source. Some other factors also influence how the viewer sees the shade and colour intensity of the triangle. Factor such as type of material determines the roughness and shininess attributes of the surface features. When there is more than one light source, such as those reflecting from other surfaces nearby, the summation of the individual reflectances can be taken to be the total intensity.

9.7.4 Texture Mapping

If the texture information for each facet is available, it may be used to fill the surface during display instead of normal colour filling. Illumination can still be applied to improve realism. The hidden line and surface removal operation needs to be applied beforehand. Since texture mapping is a raster operation, the texture array and the array of pixels indicating visible facets must be stored in parallel in the buffer memory during the operation. This operation requires powerful hardware and software because of the great deal of memory and large number of resampling operations needed.

Incorporating texture information into a simplicial network is also possible. Texture information can be provided mathematically as a function, or as an image which is typically in a raster form. Only the latter is considered here. For the raster data structure, incorporating texture information in the form of an image is quite straightforward. Incorporating texture information (known as 'texture mapping'), however, can be challenging. Knowledge of computational geometry, photogrammetry and digital image processing are needed. The texture mapping helps improve the visualization of geoinformation.

The process for texture mapping using an image involves the solving of image transformation relative to parameters of the camera used to capture the scene. If the vector data is completed with 3D coordinates, it can be transformed to match the camera orientation and then superimposed onto the image by taking into account the visibility of each vector element. The image can then be segmented by the vector elements of SNIDB, that is to say, the nodes, edges, and triangles. After the segmentation, the texture information can be stored along with each element of SNIDB. Each node has a pixel value stored as an additional attribute. For an edge, a set of pixels along the edge must be stored with the pixel size or scale factor. Eventually, this edge needs to be stretched or contracted, depending on the perspective transformation. Pixel values can be interpolated to fill the gaps between pixels (see Figure 9.26).

For a face or a triangle, all the pixels falling inside the face or triangle during superimposition must be stored. In order to normalize the camera parameters, this set of pixels may be resampled to a coordinate system that is orthogonal to the face or the triangle on the same scale. The storage of each segmentation (a face or a triangle) is then in the form of a rectangular image in which the image size is the same as the bounding rectangle of each segmentation. When the data has to be graphically viewed using different transformation parameters, the image must be re-sampled to map onto each facet. The affine transformation can be applied by taking the corners of the bounding rectangle as the control points to determine the transformation parameters. The pixels that have data values are then mapped onto the facet in a perspective view. To eliminate the gap in the resulting view, four adjacent pixels can be used as vertices of a square drawn as a quadrangle in the perspective view. The colour of the quadrangle can be determined, for example, by taking the average value of the four pixels. The process is shown in Figure 9.27.

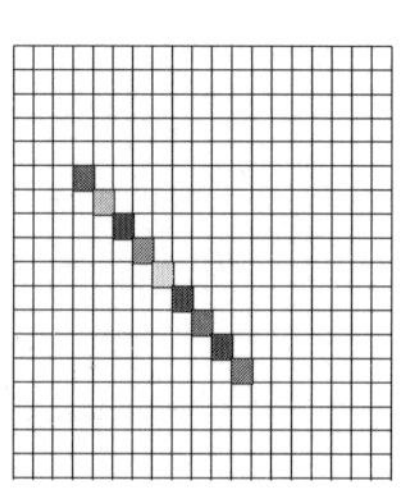

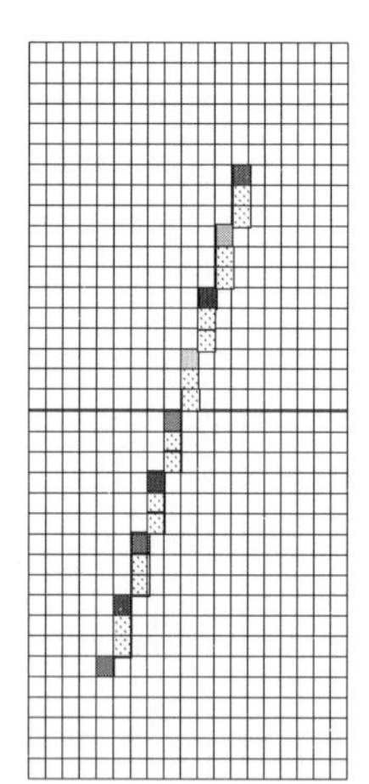

Figure 9.26 Resampling operation along an edge.

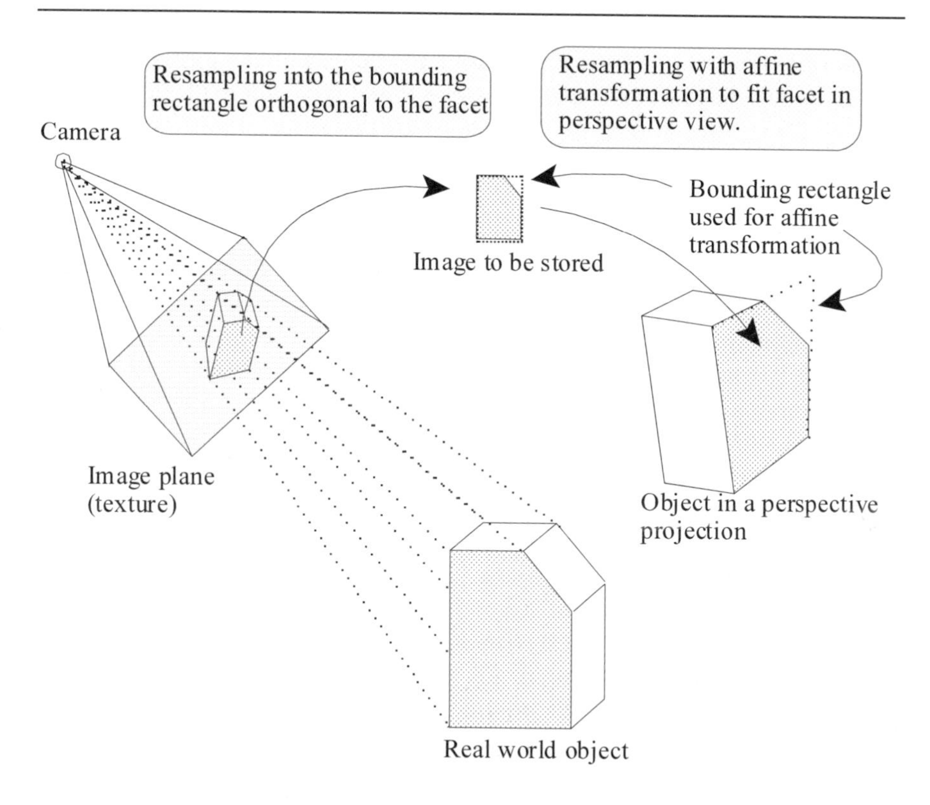

Fig. 9.27 Operation of texture mapping with respect to a face

With respect to the DBMS aspect, many modern and commercially available DBMSs (for example, Oracle Spatial, Illustra) are already capable of storing and managing an image as an attribute of a record. This capability suggests that, is feasible to incorporate the texture information into the integrated database. The data structure in a relational form may look like the following:

Face Id	Texture file	Name of texture array	Dimension (W x H)	Pixel size (mm)
347	FW1.DAT	Front_wall	30 x 50	0.2

The bounding rectangle can also be derived directly from the dimension of the texture image, using the lower-left corner of the image as the origin. The upper-right corner is just the addition of the width and height of the image to the origin, which can be started from (0, 0). The transformation problem is limited to 2D and is relative to each face. Detailed discussion about digital image transformation can be found in Wolberg (1990).

9.8 Virtual Reality

VR can be used to explore the content of information stored in an SNIDB or 3D FDS database. VR provides highly interactive, realistic and dynamic visualization. It uses almost all the visualization techniques described in section 7.5, that powerful hardware and software are needed. VR allows the continuous change of viewing position and tries to provide ways of interacting with the representation of spatial objects as it happens in reality. Understanding the spatial model can be readily achieved if the information content stored in the database is displayed appropriately. The storage of 3D coordinates and boundary representations in SNIDB and 3D FDS are compatible with many VR systems. This means VR technology can be adapted to access information stored in SNIDB directly. Thematic attributes of each feature can be translated into specific colour, shade, type of material or texture, and to graphically rendered in each scene. Spatial relationships stored in the SNIDB provide constraints for the virtual environment. For example, adjacent objects should remain close together at any viewing distance or direction in the virtual world. An experiment using the constructed 3D spatial model in the VR environment has been conducted. The 3D spatial model shown in Figure 9.17 was converted into the VRML (virtual reality modelling language) and could be viewed by many VR systems. It is expected that future 3D GIS will have a built-in VR functionality for interactive visualization and other kinds of responses, such as sound.

9.9 Discussion

The concept of a simplicial network can be applied to the integrated modelling of reality, for example the integration of terrain relief and terrain features since the problem remain an issue in geoinformation science. The spatial model resulting from this integration support operations typically needs both DTM and 2D GIS. Queries about features and relief information can occur together. A simplicial network representing the earth's surface and 2D representation of terrain objects can also be integrated into a database of 3D FDS for urban applications. For applications in the geo-sciences, for example geology, or environmental monitoring, spatial objects presented in a database of 3D FDS can be incorporated into the 3D simplicial network as constraints to facilitate better derivation of the representation of spatial objects with indeterminate spatial extent. The database in the form of a simplicial network also facilitates various kinds of 3D visualization, even when high interactivity and realism (for example virtual reality) are required, provided there is an appropriate extension of the data structure to accommodate more attributes, such as texture, or colour. Also, spatial index structures suitable for each kind of operation must be built on top of this core database for efficiency in terms of response time. Although conceptually, the simplex elements of the simplicial network data model help simplify many complex operations, the limitations on applying the SNIDB remain due to lack of 3D spatial operators that warrants further development (see Chen and Abdul-Rahman, 2006) for recent experiment on the 3D spatial operators for 3D objects in geo DBMS. These operators are metric, order and topological operators for the computation of volume, surface area, testing of containment, intersection, touch, disjoint, coincidence, and so forth. Such set operators as union, intersection, difference, or, xor, are also needed. Once these operators are available, the applicability of SNIDB will be extended significantly.

Chapter 10 THE WEB AND 3D GIS

10.1 Introduction

Recent developments in GIS are showing a general movement towards Web-enabled GIS. The gap between desktop GIS and Web GIS is closing. Applications based on network environments have already shown great potential in relation to geo-information. Examples can be online city maps and finding places (respectively routing) between points (MAP 24, 2004). Obviously, the developments in Web-enabled GIS are driven by user requirements and technology developments. But is the third dimension sufficiently exploited by Web applications?

In general, the need of 3D geo spatial data is increasing. Professionals involved in urban and landscape planning, cadastre, real estate, utility management, geology, tourism, army, etc. are especially keen on taking advantages of the third dimension. Since real world spatial objects are in 3D, it is obvious to extend GIS to the third dimension as well. However, the acceptance of 3D applications depends heavily on its profits. Therefore, we can say that the number of users could increase with the introduction of new and additional 3D functionality to the spatial system. Technologically, side, state-of-the-art computer hardware is already offering a reasonable means of dealing with the third dimension such as improved 3D visualization techniques. Among others, there are photo-realistic texturing, advanced lighting or real-time navigation that could attract more users to use such kind of applications. We firmly believe that the Web offers the possibility to make the third dimension widely accessible.

This chapter aims to provide an overview about web-oriented 3D GIS. Since we consider system architecture, data management, 3D GIS functionality and visualization (respectively user interaction) critical for Web 3D GIS, we address them in detail here. The chapter explains the needed system components and their importance with respect to the requested Web 3D GIS functionality. System architectures and possible approaches for implementing a Web-enabled 3D GIS are reviewed and explained. Directions for further research are also outlined.

10.2 Web 3D GIS

Traditionally, any geographic information system is based on the principles of data input, management, analysis and representation. Within a Web-enabled environment, these principles are represented by or implemented within the components shown in Table 10.1.

Table 10.1 GIS principles and their corresponding Web components.

GIS Principle	Web Component
Data Input	Client
Data Management	DBMS possibly extended by a spatial component
Data Analysis	GIS Library recommended on Server
Data Representation	Client/Server

In order to achieve communication between the different components in a Web environment, a Web server is needed. Since geo-data is a very specific type of data, different standards, e.g. the OpenGIS Consortium (OGC) specifications have been developed and their usage has to be taken into account (see Figure 10.1). A system composed of these components is called here Web GIS. The system should cover a complete GIS workflow within a Web environment. Figure 10.1 shows the general system architecture which is mostly "Client-Server".

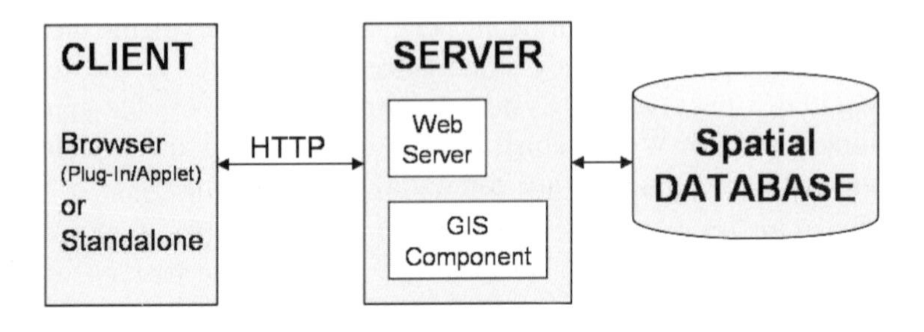

Fig. 10.1 A typical Web GIS architecture.

Figure 10.1 shows the minimum system architecture of Web GIS. The Client is an application, which can communicate with the Server through a standard Web protocol, for example HTTP. This application can either be in the form of a Web browser or stand-alone utility. In order to view and interact with GIS data, the browser needs to be extended by using an adequate Plug-In, Java Applet or both. Instead, a stand-alone application can be used, for example any GIS which supports the appropriate protocol to access other computers in the computer networks. The Web server is responsible for processing the request from the client and delivering the corresponding response. In Web GIS architecture, the Web server also communicates with the server-side GIS component. This adds spatial analysis functionality to the system. Moreover, server-side components are responsible for the connection to the spatial database, such as translating queries into SQL and creating appropriate representations to be forwarded to the server. In reality, GIS components are like software libraries, which offers special "classes" (i.e. based on object-oriented mechanism) to do spatial analysis on data.

Besides the components, a very critical aspect is the functionality offered by the client or server side within Web-GIS. Figure 10.2 shows possible distributions of functionality for a client/server system based on the concept of the visualization pipeline (OGC, 2003b).

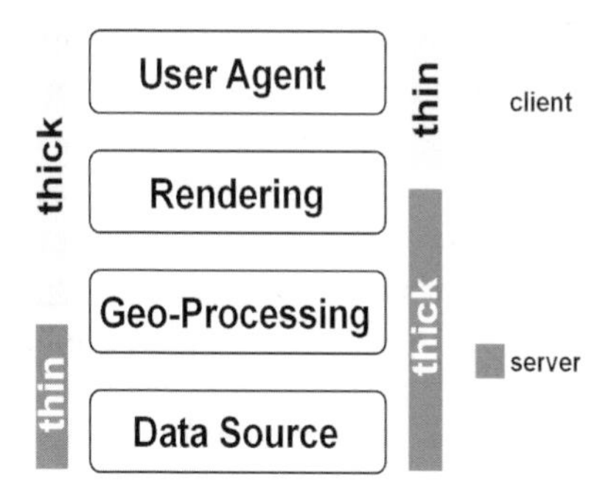

Fig. 10.2 Thin vs. thick within client server systems.

Figure 10.2 shows that a client is considered "thick" or "fat", if the main GIS functionality and the data rendering are hosted at the client side. Consequently, the server in this specific system would be called "thin". The server is called "thick" if GIS functionality and pre-rendering is hosted at server side. Within this system, the client would be called "thin". Altmaier and Kolbe (2003) exclude rendering for interactive 3D worlds on the server since real-time navigation in static images would not be possible anymore.

However, finding a balance between server and client is still an interesting question how to find the balance between server and client. Because of the system complexity, required functionality, type of application, data sets, even available funds for implementing one or another solution and user experience, no ordinary rules can be specified. The question has to be answered for each system individually. Regarding the general system architecture, 2D and 3D Web-GIS do not have many differences and the setup shown in Figure 10.1 can be used for both. Traditionally, most GIS spatial operations are very expensive and more complex compared to for example administration numerical and textural type of data. This is especially the case if the systems deal with the third dimension. Since calculations on 3D geo-information are by far more expensive than those in 2D, developers have to choose very carefully which system component is hosting certain GIS functionality. As stated before, there is no general rule. Section 10.5 discusses concrete implementations and provides answers for individual approaches. On the operational/functional side, the differences between 2D and 3D calculations are critical. The typical common operations for 2D and 3D GIS are accessing attributes or further information on objects, calculating distances and areas, buffering, routing and the nearest neighbour analysis. Whereas operations like volume calculations are limited to 3D only. Because 3D information is much more complex and has a higher quantity, the processing is much more complex and therefore takes more time and resources. 3D buffering for example needs more effort than the corresponding operation in two-dimension. These operations are done by the GIS component, either server or client side. In this respect, third party tools or an individual developed component can be used. However, there are very few available third-party tools which support 3D functionality. Therefore, the needed systems have to be customized. Individual implementations can be realized in any programming language. Here, Java language in conjunction with Servlets technology is one possibility (Vries and Stoter, 2003).

At the moment, fundamental spatial analysis, database management systems offer spatial extensions too. There are spatial extensions available for databases such as Oracle Spatial, PostgreSQL, Informix, DB2, Ingres and most recently, MySQL. Unfortunately, these software do not support 3D sufficiently (Vries and Stoter, 2003).

In order to provide the development of analysis functionality at a database level, many DBMS are supporting procedural languages as well. Oracle's

DBMS for instance offers two possibilities to create individual operations at the database level. First, there is a PL/SQL, a procedural language. Second, it has integrated its own Java Virtual Machine (JVM) in order to process Java classes at the database level. The advantage compared to external spatial analysis will mainly be in terms of a better querying performance. In addition, operators on database levels can be used by anyone who has access to the database. Therefore, basic spatial analysis operations can be reused within other applications (Jansen, 2003). Systems implementing a spatial extension are called integrated systems (Oosterom et al., 2002). Overall, the trend towards GIS in Web environments is still ongoing.

Recently, however, the term distributed GIS has been introduced. Here, a GIS will be completely distributed in a computer network. The corresponding functionality, data and certain clients operate like nodes in an object-oriented application (Peng and Tsou, 2003). However, there are no distributed GIS for the third dimension available thus far. Furthermore, since geo-data is a very specific type of data, standards have to be considered. Therefore, the Open GIS Consortium (OGC) has developed a wide range of specifications/documents which should be considered for utilization. The base for OGC-conformed GIS defines spatial data types and their relationships (the Simple and Abstract Feature Specification). Furthermore, "implementation specifications" describe interfaces and rules of exchanging or transferring data between components. In context of Web mapping, the Web Map Service Implementation Specification (OGC, 2001) has to be taken into account. It defines an interface for requesting maps. The corresponding Web Map Service (WMS) creates maps of geo-information. It has to support the operations of "GetCapabilities" and "GetMap". The operation "GetFeatureInfo" is optional but necessary for retrieving further information about objects through user interaction. "GetCapabilities" returns information about the Web Map Service itself, while "GetMap" returns the map or figure. Since editing or manipulating of data is one of the GIS principles, the Web Feature Service Implementation Specification (OGC, 2002) is a must as well. Operations of a Web Feature Service (WFS) are insert, update, delete, query and discover data. The data is represented in the form of Geographic Markup Language (GML), another OGC standard for exchanging geo-information (OGC, 2003a). Both WMS and WFS, are based on the HTTP protocol for transferring data. Among others, GML3 includes a 3D geometry and therefore suitable for Web 3D GIS. Besides, there is an implementation specification regarding 3D terrain scenes (Web Terrain Service). Altmaier and Kolbe (2003) realized that there is no specification or standard to describe the interactive 3D worlds. Therefore, they introduced the W3DS portrayal service for 3D spatial data.

OGC standards or others like the ISO/TC211 are important for the communications between components within complex GIS, especially Web-GIS. Systems can be extended easily by additional components that conform to the standards (Vries and Zlatanova, 2004). Another critical aspect is the performance of the system. If there is one bottleneck, the whole system will be affected. Therefore, system architects have to appropriately design the system. First, the base of the system should be a state-of-the-art computer hardware and appropriate applications or environments, e.g. powerful 3D visualization techniques. Due to large data amounts, data transfer between the components should be reduced to a minimum. Low band-width may cause a critical bottleneck between the client and the server. Streaming techniques, which allow data transfer in partial are popular and should be favoured for the system development. In order to achieve acceptable system performance, spatial analysis has to be done on top of a reasonable concept of storing data. Consequently, databases have to be largely employed, preferably with maintenance of topology (see section 10.3).

10.3 Management of 3D Spatial Data

In order to manage a 3D geo-information, at least the use of databases and their management systems (DBMS) are required. Object-relational modeling is the most common since relational databases are not very appropriate for storing spatial data. The object-oriented database approach faces the problem that the general acceptance and knowledge is not available so far (Connolly and Begg, 2002). The field of geo information adopts both approaches and comprises them into Object-Relational DBMS (Shekar and Chawla, 2003). As stated in section 10.2, the additional integration of spatial extensions is compulsory for GIS applications. Furthermore, because operations of 3D functionality are different from 2D, a reasonable concept of data storage is inevitable. Therefore, the two aspects of 3D geometry and 3D topology have to be taken into account. Geometry holds 3D coordinates of objects and topology holds their spatial relationships. The OGC proposes the separation between geometry and topology within databases in order to perform certain queries on geometry and topology (Oosterom *et al*, 2002). Regarding geometry, there are several DBMS available which have the ability to handle spatial data types. These data types are divided into geometric primitives of point, line and polygon. The OGC calls them simple features. However, 3D primitives like polyhedrons are missing and have to be implemented individually. Stoter and Zlatanova (2003) showed how to store a polyhedron within Oracle 9i using multiple polygons.

In contrast to geometry, the topological part is more critical. The state-of-the-art DBMS does not offer any support for 3D topology. Shi et al., (2003) and Zlatanova et al., (2004) provide a brief overview about developed topological models including additional performance tests. Oracle recently announced the integration of topology up to 4D in its database spatial extension of Oracle 10g (Lopez, 2003). The corresponding OGC specification (complex feature specification), however, is yet to be completed - in terms of the implementation specifications for complex features. Topology is the base for reasonable querying of 3D spatial data. Since there is no unique topological model available, topology has to be implemented individually. Oosterom et al., (2002) provide an overview about available approaches. A spatial data model normally meant for a certain application (Zlatanova et al., 2002a) and a generic data model for general applications is hardly available. The technique of visualization is another factor for the question in selecting a topological model. Again, there is no general rule of the selection. Topological models should fulfill tasks such as covering all possible relationship and extensibility (Oosterom et al., 2002).

Beside the geometry and topology, the spatial querying language for the third dimension poses a challenge for the database community as well. Güting (1994) concluded that in addition to SQL, a spatial query language has to provide fundamental spatial operations and reasonable ways of representing the results. Here, 3D operators on top of an ingenious data model are not available so far.

Spatial indexing is one main key to improve querying performance on geometric data - spatial objects are represented by indexes. Several different indexing methods are common but mainly R-tree, Quad-tree and P-tree are used. Furthermore, indexes are often used in conjunction with Level-of-Details (LOD) implementations (Coors, 2003; Kofler, 1998). Due to the fact that distributing within a Web environment has different requests on the volume of data, spatial objects must be as simple as possible while representing each object properly. Therefore, aspects of simplification and generalization have to be regarded as well when modeling 3D objects. Here, realistic photo-texturing is a common method to save resources. However, it has to be done patiently and it is necessary to store them efficiently. Furthermore, databases have to store the attribute information of 3D features as well.

10.4 GUI for 3D Visualization and Editing on the Web

In order to interact and communicate with information, a Graphical User Interface (GUI) has to be designed and created. A GUI is situated on top of the user agent. Because geographic information is usually very complex, this task is difficult to achieve. Moreover, the user interface is the most critical due to the fact that this is the "main gate" to the application. If a GUI is implemented poorly, an application will not be accepted by critical users. Compared to user interaction in 2D, a GUI for the third dimension is different (Cöltekin, 2002).

To develop a GUI for 3D visualization, different aspects are important. First of all, the virtual world has to be sufficient. To do so, a set of core features of creating a 3D world are needed. The technique of visualization has to cover the state-of-the-art possibilities. In the case of 3D, these techniques include are reasonable modeling of physical objects, lighting and shadowing, definition of viewpoints, and photo-realistic texturing. As soon as interaction has to be involved, using events, linking and internal/external scripting will become more important. In fact, 3D worlds including real-time interactive navigation are a requirement today. To explore virtual worlds, a user would wants to be put into the space very closely. Therefore, characteristics similar to computer games are very popular, for instance walkthrough, flying, panning and sliding. If the target is a singular object, rotating is another important real-time navigation attribute. More advanced characteristics of virtual worlds are Levels of Detail (LOD) or multi-resolution texturing implementation. Furthermore, culling algorithms should be provided in order to make sure that invisible back-faces will not be rendered. Overall, the amount of rendered polygons is a factor for the smooth navigation. Any technique which reduces the amount while keeping the world realistic should be used (Kofler, 1998).

Intuitive editing of 3D data is much more complicated than visualization. In order to provide a human readable GUI for editing, a significant amount of effort is required. This is the reason why mainly common CAD or GIS software products are used as front-ends at the moment (Stoter and Oosterom, 2002; Zlatanova et al., 2002b).

The following sections describe some of the recent developments in GUI for Web-based 3D visualization.

VRML/X3D

VRML (Virtual Reality Modeling Language) and respectively its successor X3D (Extensible 3D) were introduced by the Web3D Consortium to distribute interactive virtual worlds on the Web. Both are mark-up languages and standardized. X3D fulfills the concepts of XML. The rendering concept is mainly based on a scene graph definition and a node structure (Web3D Consortium, 2004). VRML and X3D have accomplished the basic concepts for a 3D GUI (Dykes et al., 1999); listing all the features would be too long here. Other concepts such as external authoring interface (EAI) grading techniques are also worth considering. By using the EAI, one can add individual functionality to virtual worlds. It could be developed either by using scripts or other programming languages and the 3D scenes could end up highly interactive. One good example in this aspect is accessing a database from VRML worlds to retrieve new data (Zhu et al., 2003).

Realized VRML clients in combination with HTML have already proven their ability to react as GIS user agents in many examples and prototypes (see section 10.5). However, no well-known commercial implementation is available. The most common use of VRML is within a client-side browser/plug-in implementation. Unfortunately, plug-in vendors are hesitant with shipping X3D browsers.

PHP and VRML/X3D Integration

PHP is becoming a very popular language for creating dynamic websites, particularly for generating them from databases. 2D GIS is outdated, one way is by using PHP to create database-driven VRML worlds for new 3D GIS system.

PHP can produce any text information. In PHP, one has to take control of this very information. This requires sending the VRML MIME type ("model/vrml"), and then writing the appropriate VRML nodes.

The server strips all PHP code when sending a response. So, on lines where only JSP code is present, the server simply sends blank lines back to the browser.

It is necessary to include both PHP and VRML headers, and the content type must be changed before the VRML header is set, so the final result may look like this:

```
<?php
  header ("Content-type: model/vrml");
```

```
    echo "#VRML V2.0 utf8\n";
  ?>
```

Below is an example of experiment on PHP and VRML/X3D integration for building objects.

In this experiment, one of the objectives is to display a 3D building in VRML code using a database. Figure 10.3 shows the VRML file containing some building datasets.

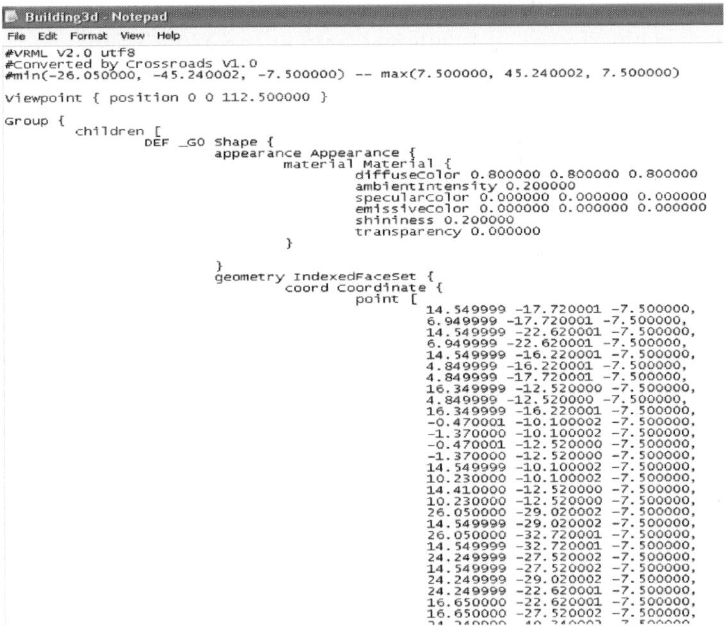

```
Building3d - Notepad
File  Edit  Format  View  Help
#VRML V2.0 utf8
#Converted by Crossroads V1.0
#min(-26.050000, -45.240002, -7.500000) -- max(7.500000, 45.240002, 7.500000)

Viewpoint { position 0 0 112.500000 }

Group {
        children [
                DEF _GO Shape {
                        appearance Appearance {
                                material Material {
                                        diffuseColor 0.800000 0.800000 0.800000
                                        ambientIntensity 0.200000
                                        specularColor 0.000000 0.000000 0.000000
                                        emissiveColor 0.000000 0.000000 0.000000
                                        shininess 0.200000
                                        transparency 0.000000
                                }

                        }
                        geometry IndexedFaceSet {
                                coord Coordinate {
                                        point [
                                                14.549999 -17.720001 -7.500000,
                                                6.949999 -17.720001 -7.500000,
                                                14.549999 -22.620001 -7.500000,
                                                6.949999 -22.620001 -7.500000,
                                                14.549999 -16.220001 -7.500000,
                                                4.849999 -16.220001 -7.500000,
                                                4.849999 -17.720001 -7.500000,
                                                16.349999 -12.520000 -7.500000,
                                                4.849999 -12.520000 -7.500000,
                                                16.349999 -16.220001 -7.500000,
                                                -0.470001 -10.100002 -7.500000,
                                                -1.370000 -10.100002 -7.500000,
                                                -0.470001 -12.520000 -7.500000,
                                                -1.370000 -12.520000 -7.500000,
                                                14.549999 -10.100002 -7.500000,
                                                10.230000 -10.100002 -7.500000,
                                                14.410000 -12.520000 -7.500000,
                                                10.230000 -12.520000 -7.500000,
                                                26.050000 -29.020002 -7.500000,
                                                14.549999 -29.020002 -7.500000,
                                                26.050000 -32.720001 -7.500000,
                                                14.549999 -32.720001 -7.500000,
                                                24.249999 -27.520002 -7.500000,
                                                14.549999 -27.520002 -7.500000,
                                                24.249999 -29.020002 -7.500000,
                                                24.249999 -22.620001 -7.500000,
                                                16.650000 -22.620001 -7.500000,
                                                16.650000 -27.520002 -7.500000,
```

Fig. 10.3 3D building's VRML file.

PHP scripting is added to this shape, which lets us use dynamic data to change the sphere's position in space (translation X Y Z), its color (*diffuseColor R G B*), and its radius.

The Prototype and Process details

The basic idea of a prototype is to organize 3D geo spatial objects in a DBMS and to query them via an Internet browser. Geo spatial objects contain both spatial and non-spatial (administrative) information. The spatial information can be visualized after conversion into VRML or X3D and the non-spatial attribute information can be presented in (dynamic) HTML pages.

Figure 10.4 shows a standard request process—the page is requested via a browser. The request calls the designated PHP, which interacts with a database. The model given below explains the prototypical process.

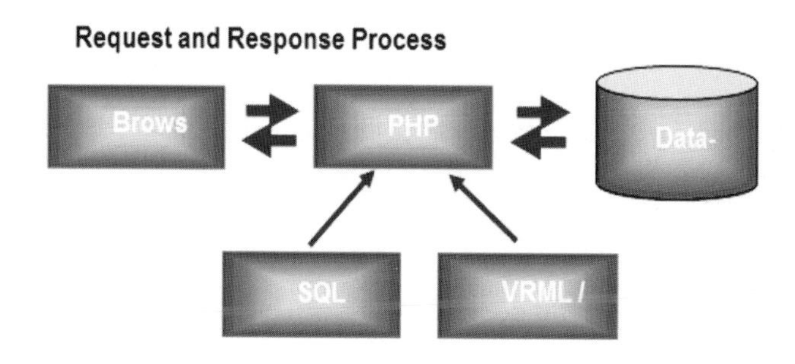

Fig. 10.4 The request and response processes.

After receiving a response, the system follows the flow shown in Figure 10.3. The database sends the requested data to the PHP, which formats the data and sends the response to the requesting browser. In our case study, the data is returned to the PHP, which generates a VRML scene using the data from the database.

On a client request, a connection is made to the DBMS and the spatial information of interest is selected from the DBMS and converted into X3D/VRML. A browser plug-in at the client side makes it possible to view the VRML or X3D output. VRML and X3D provide the possibility to start a script when a user clicks on an object. This functionality is used to retrieve the non-spatial information that is linked to a 3D geo-object. Via the VRML/X3D plug-in, a request is sent to an application server. The server receives and interprets the incoming information and sends a HTML with the required information back to the browser.

For retrieving the spatial and non-spatial information from the DBMS, a technique is needed to communicate between a client and a database on a server. For this communication, several techniques are available such as ColdFusion, ASP.NET, ASP, JSP or PHP. The choice of the used technique is dependent on the Web server used.
The detailed architecture of publishing a 3D dataset on the Web is shown in the following Figure 10.5.

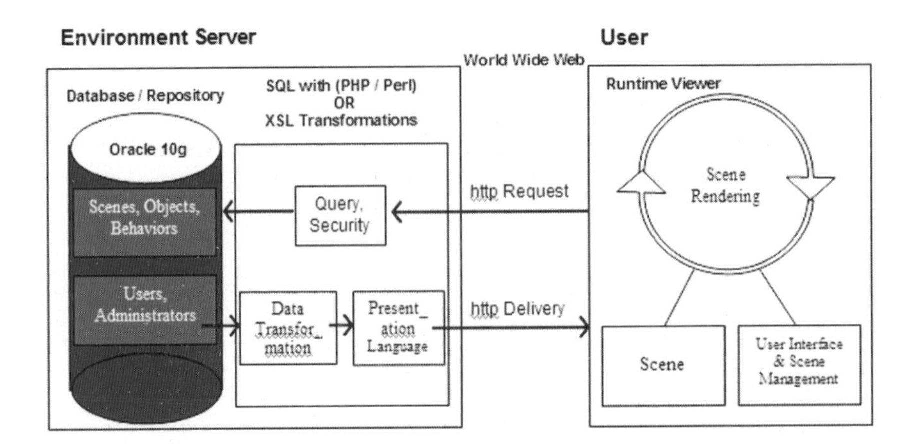

Fig. 10.5 Web publishing architecture of 3D datasets using 3D-GeoDBMS.

To show the possibility to query 3D geo spatial objects via an Internet client, first a simple prototype was built, based on Microsoft technology.

The following section describes (in detail) how the dynamic Web 3D visualization could be performed.

An Approach Towards Supporting Dynamic Scene Generation

Today, the dynamic generation of HTML pages is a standard functionality of all commercial database systems. This feature has been proven to be a very effective and practical approach to support the (two-dimensional) visualization of information stored within database systems. In this section, we outline how a similar functionality can be realized to dynamically generate VRML scenes from a database management system (DBMS). This approach overcomes many of the limitations of static VRML scenes, by exploiting the persistence, scalability and security mechanisms of database management systems. In addition, it also provides a direct way to efficiently generate three dimensional visualizations from existing information in the database.

Here, a "Geometry" data type allows, for example, to store all apartments of a building as VRML scenes together with the walls and its floor information along with its ID. It is possible to select a subset of all apartments with certain properties and merge them in a new scene, e.g., in order to display all apartments that have already been rented out to tenants.

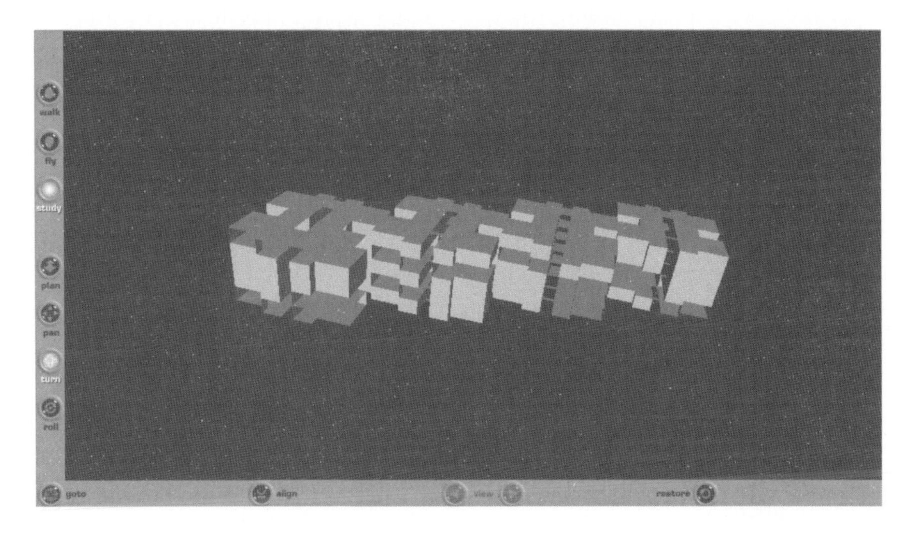

Fig. 10.6 A 3D dynamic data visualization of 3D building data on the Web.

Figure 10.6 shows a complete building structure generated using VRML client application installed as a plug-in with Web browser. The figure also shows the complexity of the 3D data and generating a dynamic scene through database onto the Web browser as indicated in Figure 10.5.

Dynamic scene generation in VRML/X3D using the data from the Oracle Spatial 10*g* was performed in a way given below:

A database connection is used as a statement inside PHP coding. Here are the strings (i.e. scripts coding) used to connect to the Oracle database:

```
<?
$connection = OCILogon (“User Name”, “Password”, “service name”)
              or die (“cannot connect to database”);
?>
```

The SQL in this case is quite simple — return all data contained in the table after execution:

```
<?
  $stmt = OCIParse($connection, “select * from building” )
     or die (“cannot select”);

  OCIExecute($stmt, OCI_DEFAULT);
?>
```

Once the result is assembled, the code loops through all of the records and displays as many floors (of the building) available in the database. This is the loop:

```
<?
  $count = 0;
  While (OCIFetch ($stmt))
  {
        color = ociresult ($s, "color");
        radius = ociresult ($s, "radius");
  ?>

  DEF Floor <? $count ?> Transform
  {
        translation <? $count*15 ?> 0 0
        children
        [
        Shape
  {
        appearance Appearance
  {
        material Material
  {
        diffuseColor <? $color ?>
  }
        }
        geometry Sphere
  {
        radius <?$radius?>

        } } ] }

<?
count++;
}
```

Notice how the values from the database are inserted into the VRML. Two strings, colour and radius, are set to the values from the database and then displayed in the VRML code as *<? $color ?> and <? $radius ?>*.

The translation is handled by an integer (count), which keeps track of the number of records and displaces the floor by 15 on the x-axis iteration of the loop.

The variable count has another use as we generate objects, we give them a name: DEF Floor1, DEF Floor2, and so on. We do this by inserting the count into the VRML node's definition DEF Floor <? $count ?>.

A DBMS supporting three-dimensional visualization must be able to generate new VRML scenes, both from existing operational business data represented by conventional data types and from existing multimedia data represented by specialized media data types. Within the building, statistical data about how many persons can live in one apartment could be visualized by an arrow diagram, where the arrow size is proportional to the number of family members allowed to live in an apartment. As always, free the database resources at the end of the operation.

```
<?
   OCIFreeStatement ($stmt);
    OCILogoff ($connection);
?>
```

The VRML client has to be able to directly read and write the DBMS from within a VRML scene. For more advanced interaction modes in multi-user environments, this mechanism needs to be complemented by an event handling system. This allows signaling a change in a scene to all other users actively working on the same scene. The VRML event handling of the other users can then react by appropriately updating the scene, e.g. by reloading a part or the whole scene.

The most important benefit of the above-outlined approach is that by means of storing VRML scenes within a DBMS, we achieve persistence of changes to scenes. Furthermore, the multi-user access control enables the sharing of VRML data among multiple users, thus we move from isolated, static scenes to shared spaces of dynamically generated three-dimensionally (3D) visualized information. Scalability is achieved by loading and generating scenes and scene components dynamically either at loading time or at run time. The corresponding loading/generation schemes can be determined both by physical characteristics of the VRML scene and the logical structure of the application. Controlled access to scene data is supported by the security and view mechanisms of the underlying DBMS. For example, in the "Building" scenario, one might display the rent of an apartment only to authorized members or customers but not to visitors.

Java3D

Another instrument for creating 3D world on the Web is Java 3D. The Java3D library is a freely available API for developing Virtual Worlds in Java (Sun Microsystems, 2004). Therefore Java3D classes can be used by Java Applets within HTML pages. Java3D's functionality is almost the same with that of VRML and X3D. Savarese (2003) introduces them briefly. One big advantage compared to plug-in based solutions is that developers have more control about rendering and user interaction. Another is the transformability. Compiled Java3D classes can either be used as standalone application or applet. In contrast to the mark-up languages of VRML or X3D, Java3D requires much more programming knowledge (Diehl, 2001). This is probably one reason why only few solutions have been realized using Java3D. One example for implementing Java3D within a geo-related application is the DEMViewer (see Taddei, 2003).

10.5 Current and Possible Approaches in Urban Planning

The steady growth of urban environment worldwide poses challenges to our society. In order to avoid chaos and confusion, urban scenarios like cities and their complex streams have to be planned well. Therefore, geo-information and corresponding spatial data must be able to support planners and their decision makers tremendously (Laurini, 2001). Possible fields of applications are listed in Table 10.2.

Table 10.2 Possible fields of applications within urban environments for Web-based 3D GIS (after Altmaier and Kolbe, 2003).

Sector	Description	Example
Event management	Simulation of the event to attract people	Offering the possible 3D view of a certain seat in a stadium
Facility management	Management of big building complexes	Organizing the room availability of a hospital
Navigation support	Car and pedestrian navigation systems	Location-based service displaying the recent position and its environment
Environment	Environmental Topics in Cities: noise characteris-	Visualizing the emission

	tics, air flows, emission dispersions, etc.	dissemination
Disaster/emergency	Organizing the work-flows in the case of an emergency	Directing rescue teams through complicated envi-ronments with support of real-time data
Supply engineering	Management of supply related tasks	Organizing the power net-work

Table 10.2 shows many possible useful scenarios for 3D applications in an urban environment. While some of them, for example, event management mainly deals with visualization only, there are applications involving spatial analysis, particularly, the topic around disaster and emergency management.

However, recent 3D GIS implementations mainly cover 3D visualization and simple interactive components like accessing additional information. Other general GIS principles such as data analysis are still missing. The reason for this is that the related data management is not suitable for real 3D functionality (Nebiker, 2003). However, there are a couple of prototypes available which point towards the real 3D GIS. The following brief descriptions introduce browser-based and stand-alone front-ends.

10.6 Realized Browser-based Solutions

As stated in section 10.2, browser based solutions are almost represented by some kind of browser plus plug-in approaches. The following examples use mostly HTML-based Web pages which have VRML embedded files.

A prototype system of 3D GIS (Zlatanova, 2000)

The developed system is a typical example of a very thin client, i.e. based on HTTP, CGI scripting (realized in Perl), VRML and HTML documents which are created on-the-fly. The VRML delivers the 3D graphics information obtained as a result of spatial queries or/and provides means to query graphically the objects observed in the 3D scene (by standard VRML nodes). HTML documents are used to visualise text and images, to specify SQL queries, or introduce new values for edited elements. Web and VR browsers on the client stations are used to interact with the 3D model(s) and specify queries. The data are structured according to the

Simplified Spatial data Model (SSM) topological model and maintained in a Relational Database Management System (RDBMS), namely MySQL.

Requesting information about a particular object can be done either by typing its ID in a HTML form or by clicking on the corresponding object in VRML (its graphical representation). For example, a click with the mouse on a building activates a CGI script, which delivers a "Query-Result" section (HTML). The user selects the needed information from a "pull-down" menu that is created on-the-fly with all the information available for the object in the database.

Extracting a group of objects according to a criterion is completed by directly typing SQL query at the "Query" section. The result of the query is displayed either in an HTML or in a VRML document. These documents are created on-the-fly only with the information related to the objects of interest. The same mechanism is used to create DELETE, UPDATE, and INSERT forms to edit data. The free access to the database provides a mechanism to specify and display a wide range of spatial queries. Examples of such queries are "Which is the highest building?", "Show the buildings in a particular area", "Show all streets", "Show all administrative buildings".

An advantage of the system is that clients practically do not use any specific software besides a Web browser and a VR plug-in. The system also does not have a specific GIS component since the SQL queries are directly sent to the database. The spatial functionality is provided by operations at database level. The major disadvantage is eventual overload of the server in case of too many users. The performance of the system has not been tested for multi-user access. Another disadvantage is increased complexity of the VRML file if elaborated point-and-click operations are needed. To be able to work with freeware VR browsers, all the interactions with objects are incorporated in the VRML (using special VRML nodes). Therefore, in many cases, the size of the VRML file can increase drastically.

GOOVI 3D (Coors and Jung, 1998)

The system architecture is a medium client-server where most of the functionality is provided at the server side but some functionality is also kept at the client side. The components of the system are VRML, HTML, Java and warehouse. The warehouse consists of files organised on the server. The interface to the data warehouse is done by COBRA IDL and is based on IIOP protocol. The two kinds of queries, i.e. obtaining additional information about a selected object and extracting several objects as a result of specific queries, are also implemented. In the first case, this is done by attaching to the objects in the VRML files hyperlinks to a HTML page (the pages are stored in the warehouse) or, more dynamically, by Java script nodes. In the second kind of data queries (objects which meet specific conditions), the server has to access at database level in order to perform the queries. The results are represented by highlighting the objects of interest in the current VRML scene using Java and IIOP protocol. Thus no new VRML file is created. Since the system is invented for discussing urban plans, editing/modification operations are not implemented. The authors make a suggestion for SQL node in VRML that can be directly used to connect to DBMS and extract information. First implementations of the system use RDBMS to store objects as VRML nodes and information about them as HTML pages. Later implementations made use of more generic representations in Oracle, using the topological model UDM (Coors, 2003). The advantage of the system is that it is a relatively thin client-server system, allowing implementations without large resources at the client side. Part of the functionality (data query) is performed at the server but highlighting of the objects of interest is at the client side. In this respect, the system is better balanced than the previous one. The system however is a bit dependent on the file organisation in the warehouse (i.e. mixture between files and DBMS storage). The major disadvantage is that the extended protocol IIOP is used (which is generally not available).

SALIX (Lammeren and Hoogerwerf, 2003; Wachowicz *et al.,* 2002)

SALIX is a typical example of a thick client. The system is intended for interactive landscape planning, i.e. planning trees and bushes and simulating their growth. The GUI is based on the Cortona environment, using VRML and Java to provide all the functionalities. DBMS is used only to store the objects of interest (a variety of tree and bush species). The objects are manually placed in the field of view. A large number of toolbars give the users the possibility to inspect certain constraints, the distance between the planted trees in different stages of their life, to simulate growth, to create conglomerates of objects from the same type, etc. The significant aspect of this system is the extended functionality in terms of interactions and manipulations. There are still more improvements necessary toward making real use of functionality available at DBMS (currently used only for object storage).

Accessing Geo-DBMS Using Web Technologies (Vries and Stoter, 2003)

Vries and Stoter (2003) describe two prototypes using a web environment to query 3D spatial data and their attributes. The implemented applications focus on reasonable ways to visualize query results within a web browser. Because the operations are hosted on the server-side, the system is represented by a thin client and thick server. The realized prototypes can be differentiated by the following technologies.

- **VRML and Microsoft-specific technologies**

  This implementation uses common Web technologies to achieve a 3D GIS. Geo spatial data is already available within VRML files, and its attributes can be queried dynamically. These are stored in Microsoft's Access database system. Active Sever Pages (ASP) technology combined with the Internet Information Server (IIS) as Web server environment is used to offer interaction with the database. The served VRML world is embedded within the main frame of the HTML Page. User interaction is possible in form of querying each objects attribute data. If the user clicks on an object in the VRML world, a request is sent to the server. After connecting to the database, ASP creates an appropriate HTML fragment which holds the requested attribute data in a table and embedded in the second frame of the application. This approach is vendor specific and it only works properly on Microsoft (MS) components.

- **X3D, Java Servlets, XSQL and Oracle 9*i***

  This prototype system is based on an integrated database architecture. The underlying DBMS hosts 3D spatial data as well as their attributes. Oracle 9*i* and its spatial extension are preferred. Server-side, the system is based on a Java Servlet Container, like Apache Tomcat, and the Apache HTTP server. In detail, the prototype is using XML specific Java libraries to query (XSQL) and exchange data. The libraries are part of Oracle's XML Developer Kit's (XDK) which are integrated in Tomcat. Among the XSQL servlet and others, the XDK provides a XML parser and XSLT processor. In order to visualize the queries, the XML response of the database is transformed to X3D using XSLT style sheets on-the-fly. On the client side, the browser window is separated into three frames. The main frame is for showing the virtual world, another for displaying the object's corresponding attributes using HTML tables. The third frame offers HTML forms in order to query the database for spatial objects. Once a query is performed, the main frame will visualize the new scene.

This state-of-the-art implementation demonstrates nicely the advantages of a fully XML based system. Furthermore, it can be integrated into any platform which supports Java programming language. Figure 10.7 shows the prototype's client interface.

Fig. 10.7 A prototype system using Web technologies to access Geo-DBMS.

Pilot 3D of the GDI NRW

The Special Interest Group (SIG) 3D of the Geo-Data Infrastructure North-Rhine Westphalia, Germany (GDI NRW) has proposed their first prototype (see, Groger et al., 2004). The 3D city model is based on the geometrical objects point, line, surface and body and has been presented in Gröger *et al.*, (2004). The corresponding application logic - realized in Java programming language - offers a standard based (OGC and ISO19107) solution to visualize 3D urban data. The proposed data model is used in 3D city models, virtual flights and other projects which are able to improve planning processes. For interactive 3D visualizations, VRML is currently used. A first published result has been presented by the SIG 3D and is available online (SIG 3D, 2004).

Overall, the "Pilot 3D" project can be seen as a prototype scenario in order to prove the value of a standard-based Spatial Data Infrastructure. The most important fact is that the SIG 3D proposes their own extension of the Web Terrain Service called Web 3D Service (W3DS).

10.7 Stand-alone Solutions/Toolkits/Front-ends

Most CAD or GIS can be integrated into a Web environment. They can be used as a user agent on the client. Stoter and Zlatanova (2003) describe the approaches using ESRI's ArcScene and Bentley's GeoGrapics iSpatial to visualize and edit data. These examples do not cover the integration into a Web environment. Nevertheless, one can do so, because other software products are difficult to use and they are not very suitable for inexperienced users. Therefore, different institution or companies have created special 3D applications. Geonova's Digital Landscape Server (DILAS) product line is one promising approach. The following descriptions illusttare briefly the application and its components.

DILAS 3D (Nebiker, 2003)

Geonova's commercial product line DILAS offers a large variety of modules which can be seen as 3D Web-GIS. The DILAS server and manager are the main components of the system. They are responsible for characteristics like data storage, management, representation and scene reconstruction. The DILAS modeler is an extension on Bentley's Microstation V8. This component integrates the creation and edition of new 3D objects and their corresponding styles. Moreover, the modeler benefits from the possibilities of Microstation due to the fact that it uses its Java API. In order to publish 3D worlds on the web, the DILAS scene generator is the key component. In conjunction with the visualization product G-VISTA, it can generate complex 3D scenes like city models. These can be served by any web server. Most recently, Geonova announced the new OGC conform Web Map Service. Therefore any client which is supports this specification can be used (GEONOVA, 2003).

The whole concept and the already implemented features look very promising for the use in urban planning. Based on a state-of-the-art object-relational DBMS, DILAS offers managing, editing, reconstructing/visualizing and publishing virtual worlds. However, editing and managing of 3D scenes is only possible within an intranet network. Furthermore, there is no 3D functionality offered by default. Nevertheless, the examples shown are impressive.

GIERS (Kwan and Lee, 2003)

Kwan and Lee (2003) describe a developed GIS-based intelligent emergency response system (GIERS) which implements 3D routing features up to the inside of buildings for rescue teams in real-time. The results is a navigable 3D GIS which includes building internal navigation as well as associated ground transportation possibilities of a city. The underlying 3D data concept comprises a topological node-relation structure which is used for routing operations and worked within a relational database model. On the technological side of the implementation, mainly Microsoft specific technologies are used. Furthermore, depending to its purpose, the system is able to communicate with mobile devices as well as through the Internet (Kwan and Lee, 2003).

10.8 Summary

This chapter introduces the complexity of 3D GIS on the Web. System architecture, data management and GUI visualization are seen as critical aspects. The chapter also discusses VRML/X3D and Java3D as visualization techniques for distributing virtual worlds on the Web.

Due to the fact that there are not many core systems available, research on 3D Web-GIS needs a lot more effort to be successful, especially in the field of data management which lacks reasonable approaches. Furthermore, data updating (including "dynamic" updating) should be addressed as well. Finally, 3D Web services functionality has to be made available for the next generation of 3D GIS system.

Chapter 11 CONCLUSION AND FURTHER OUTLOOK

Research and development within the scope of 3D GIS are now extensive. This book only deals with some parts of it. The emphasis here has been on the conceptual and logical design of a 3D spatial model and how it can be constructed and applied. Some examples utilizing the 3D spatial model with respect to the design introduced in this book are also given. This final chapter concludes the discussion drawn from this research and suggest some recommendations for further research.

11.1 Summary

Several problems associated with 3D GIS were identified in chapter 1. The scope of this book, however, restricts the emphasis to various stages of the design and construction of a 3D vector spatial model. This kind of model permits the integration in one database of two kinds of real world objects: determinate and indeterminate spatial objects and their components. This integration permits better representation of the spatial relationships between the two types of spatial objects. Determinate spatial objects - objects with discernible boundaries, like buildings and roads - can be represented directly by the elements of the model. Indeterminate spatial objects - objects with indiscernible boundaries, like soil strata, temperature, and mineral deposits - require indirect representation. Given a specific type of property and a given property value, or the property range, the boundary of an indeterminate spatial object can be derived from the surrounding neighbours. In a vector spatial model, the neighbours may be represented as a point, line, surface, or body feature.

When the boundary of an indeterminate spatial object has been derived, this object then becomes a determinate object capable of being visualized and allowing further spatial analysis (computation of volume, surface area, relationships with other spatial objects, etc). To make this possible, the neighbours must form a spatial unit permitting the performance of operations (interpolation, classification) to make the boundary of the indeterminate spatial object explicit. To obtain an accurate result, the (derived) boundary of the indeterminate spatial object and the characteristics of the neighbours must be taken into account as constraints. For example, underground discontinuities like geological faults, obtained from interpreting seismic data, may have to be incorporated directly into the spatial model so that the derivation of orebody from drillhole samples of mineral deposits can be obtained more accurately.

The review of the current situation indicates that existing systems do not provide adequate 3D modelling tools for earth science applications needing to model the relationships between determinate and indeterminate spatial objects. Moreover, the components of spatial objects are often represented in separate spatial models, such as in models of terrain relief in DTM and terrain features in typical 2D GIS. The consequence of these is the difficulty of accurate representation of the relationships between objects in a spatial model. It is evident that the key problem is the lack of a spatial data structure suitable for this kind of modelling which also permits the adaptation of various available technological developments to be implemented as functions of 3D GIS. Therefore, an appropriate spatial data model has to be developed to make it possible to derive such a spatial data structure. Although attempts towards the design of a 3D spatial data model have been made, this aspect of integrated modelling of the two types of spatial objects had not been adequately addressed. The main objective of the research was, therefore, to design a data model suitable for the integrated modelling of the two kinds of spatial objects and accordingly to suggest a simple method of constructing a spatial model as well as to demonstrating its usage.

The study commences with a review of all the necessary fundamental concepts of geo-spatial modelling incorporated into the design, construction and maintenance phases. Although different theories and concepts abound, only those supporting the design of a spatial model are reviewed and ordered with respect to the conceptual and logical design phases. By relating and bringing some order into those theories and concepts, the study also contributes to the further development of spatial theory.

Conceptual design

Since a 3D GIS needs to adapt various technological developments for its functionality, a review of these technological developments was carried out. In addition to this, the present architecture of the geoinformation systems and future development trends are analysed and differentiated into four stages of evolution: independent subsystems, functional integration, client/server, and structural integration. The independent subsystem is the common approach, since a GIS has been developed whereby available subsystems in the form of hardware and software are taken as components of a GIS. These subsystems evolve into software modules of a GIS in the next evolution stage, the functional integration. However, few of the GISs developed at this stage can provide all the functions the users require. The client/server architecture, which is evolution stage three, emerges offering an intermediate solution. This architecture makes use of communication technology to exchange information between independent subsystems connected on-line. Nevertheless, the architecture of the systems in these three evolution stages still relies on various independent data structures specific to functions or subsystems.

Since separate data storage requires different DBMSs, many problems persist. These problems are summarized in chapter 3. The monograph anticipates evolution stage four, the structural integration, that is expected to offer solutions with all functions relying on a common database. This database provides the information necessary for all the operations in geospatial modelling. A unified data structure is the basis of the system. The system provides various database views and spatial index structures specific to functions or operations on top of the unified data structure. The client/server approach can be adopted on top of the structural integration architecture which allows each developer to concentrate on a set of functions of 3D GIS. The review of some attempts towards structural integration with respect to 3D GIS shows that the design of spatial data model is needed to permit the derivation of a unified data structure for a 3D GIS employing a structural integration architecture to accommodate both direct and indirect representations of spatial objects.

To contribute to the development of 3D GIS adopting architecture based on structural integration, the design of an integrated data model and the development of the method to construct the spatial model were carried out. The simplicial network data model (SNDM) is the result of the conceptual design. The SNDM provides general concepts valid for spatial models ranging from 2.5D to nD. The SNDM has the following properties:

1) Theoretical aspects:

- i) The components of the model are distinguished into geometric, feature and thematic class levels in the same way as FDS.
- ii) Complex spatial objects are decomposed into simplices. The Delaunay concept of taking complex objects as constraints is used in the decomposition method to provide spatial units suitable for the indirect representation. All simplices contribute to the geometry of the simplicial network.
- iii) A simplicial network as well as its components can be described using graph the theory. Each simplex is a complete

graph, therefore, a simplicial network is a network of complete graphs with different degrees of nodes. This concept makes the simplicial network a sound and consistent structure. Each network has mathematical characteristics that accord with a generalized Euler equality.

2) Practical aspects:

i) The network provides basic computation units suitable for finite element analysis.
ii) The network accommodates both direct and indirect representations. The direct representation implies a high fidelity property of representing spatial objects.
iii) The network has a locality property, so it is suitable for use as a structure for the storage of a large database where a spatial model can be maintained without large perturbations to the model as a whole.
iv) The irregularity of the network makes it adaptable to spatial variation in reality. This makes the spatial model versatile.
v) The network is a complete tessellation of space, allowing more freedom to navigate within the spatial model using various means, such as topology, order, metric computation or their combinations. For example, the derivation of iso-lines or iso-surfaces makes use of a combination of different means to navigate in the spatial model, while query about features make use of topology as a navigation means.

Each component of a simplicial network has the following properties:

i) convex shape
ii) irregular shape
iii) finiteness, therefore, it is verifiable against a complete graph
iv) simplest geometry in its internal dimension.

These properties make it possible to automate many operations ranging from the construction of a 3D spatial model (for example, the constrained Delaunay network formation) to the derivation of information as required by applications in earth sciences (for example, the computation of spatial gradient, iso-lines, iso-surfaces) as indicated in chapter 1.

Logical design

With respect to the logical design, a unified data structure (UNS) can be derived from the SNDM. This allows for the handling of a spatial model by a single DBMS. Two different logical designs using the relational and object-oriented approaches ensure that a SNDM is feasible. The UNS provides elements for storing the necessary information for various operations. Regardless of the speed of the spatial operations, a relational UNS can be implemented using many commercially available DBMSs which already provide the basic operations to create, retrieve and update a database and its elements. Normalization using Smith's method is applied to obtain relational UNS, providing better updating the database. Typical relational DBMSs do not provide functions for spatial operations. These operations have to be implemented in addition to the basic database operations.

The logical design using the object-oriented approach shows that SNDM can be implemented differently for better performance that is not well provided for by the relational approach as a result of the unsuitable indexing method. The implementation using C++ in the ISNAP program demonstrates the practicability of the object-oriented approach. Instead of relating components of a spatial model by joint operations and Cartesian products, as is typical in the relational approach, relationships among the components of a spatial model can be implemented as pointers. Spatial searches can be more efficient, as can be seen in operations like the derivation of grid DTM and contour lines.

Construction of a 3D spatial model

For practical use, a method of constructing such a spatial model must be available. The constrained network construction is the method of constructing a spatial model based on SNDM. Incorporating representations of determinate spatial objects as constraints into the simplicial network is the most important issue. For 2D network construction, both raster and vector approaches are available and ready for use. Since no simple method for 3D network construction with constraints is available, generalizing a 2D algorithm for 3D is feasible. The vector approach is, however, very complicated to generalize for 3D network construction. Generalizing the raster approach was achieved within this study. A general method valid for nD is devised to incorporate constraints into the simplicial network. The method is based on the invariant property of Voronoi regions under Voronoi tessellation, using distance transformation. The geometry of line and surface features can be embedded within the 3D simplicial network as required. A further achievement is the construction of a spatial model based on 3D FDS required as an intermediate structure for storing features to be used as constraints in the 3D network construction. This achievement is, however, limited to man made objects such as buildings and roads extracted by photogrammetric digitizing from a stereo model. The construction of this kind of model is achieved by using the digitized outline of the building roofs and DTM, as explained in chapter 8.

Implementation

The object-oriented UNS and method for constrained network construction were implemented in the ISNAP program. Some functions which are only available on separate systems (2D GIS and DTM) can now be implemented in ISNAP, together with some additional functions required for 3D modelling. In short, ISNAP has the following functionality:

- 2D Delaunay network construction with constraints
- 3D Delaunay network construction
- Graphic display:
  - orthogonal, perspective and stereo views
  - wireframe or surface illumination
  - hidden line and surface removal
- Query of point, line and surface features that can be performed in any display view
- Derivation of contour lines, contour surfaces
- Derivation of regular-grid DTM.

Apart from ISNAP, other developments in the field of 3D GIS have been carried out. TREVIS has been developed to explore the capability of the relational approach and 3D FDS for 3D GIS and uses as tool for 3D visualization with some 3D editing capability. TREVIS can perform various kinds of queries using functions provided by a commercial relational DBMS, dBASE IV.

Testings

The applicability of the SNDM is demonstrated through three tests. These tests were conducted using both TREVIS and ISNAP as tools for constructing the model: query, process, and visualization. The first test is specific to the problem of the integrated modelling of terrain relief and terrain features typically handled separately by DTM and 2D GIS. ISNAP was used to perform constrained triangulation, overlaying, query of features, deriving contour lines and regular-grid. Basic GIS and DTM functions could be performed on one database. It can be concluded that the integrated modelling of terrain relief and terrain features is achievable using SNDM.

The second test shows that the representation of a surface in the form of a simplicial network can be integrated into a spatial model in the form of a 3D FDS that contains the representation of 3D spatial objects. The study area was the central area of Enschede, The Netherlands consisting of different kinds of buildings. The terrain of the study area was represented in the form of simplicial networks that facilitate the construction of the geometry of 3D representation of buildings. Representations of the footprints of these buildings were incorporated into a simplicial network of the surface, successfully integrated with 3D objects representing buildings and stored within a 3D FDS database in the subsequent process.

The third test demonstrates the construction of a 3D simplicial network and the derivation of the boundary of indeterminate spatial objects from this network. A tetrahedral network was constructed from simulated borehole data using the ISNAP program. Iso-surfaces that are the assumed boundaries of a soil layer (for instance, clay) were derived from this tetrahedral network and then visualized in perspective and stereo mode. Determinate objects such as buildings and roads could be incorporated into this network for visualization purposes, or for complex analysis.

The test results shown in chapter 9 are obtained from TREVIS and ISNAP. This ensures SNDM can be implemented and the resulting spatial model is practicable. The model can fulfil various requirements, that is integrated modelling of determinate and indeterminate spatial objects, supporting complex queries, various kinds of visualizations. Boundaries of indeterminate spatial objects can be derived and visualized.

The ISNAP program demonstrates that the 3D spatial model based on UNS can facilitate various operations. Tasks in the scope of 3D GIS (described as the functionality of ISNAP) that typically require many different systems and databases can be carried out using ISNAP and an integrated database. This makes the ISNAP a simple example of 3D GIS adopting the structural integration architecture. Various functions are available and reachable from one control panel with a common user-interface, making the system more convenient to use and significantly reducing time in dealing with many different systems. Since one database can facilitate many operations, data redundancy due to storing duplicate data in different databases (for example databases of terrain relief and terrain features) is eliminated. Accessibility to each data element is improved because all components of spatial model are stored in one database. Users need not deal with uncertainty during spatial query or computation. This means many requirements stated in chapter 3 can now be fulfilled. ISNAP also demonstrates that various technological developments, construction of spatial model, query and deriving information from the model, 2D and various 3D visualization techniques and so forth, can be integrated into one system that uses SNDM.

11.2 Further Research

In this book, a simplicial network data model has been described. Previous chapters have shown the design and implementation of the unified data structure and method to construct the corresponding spatial model. Some of the practicability of the constructed model has also been demonstrated. However, further investigations and developments still need to be carried out as listed below:

- Development of tools for 3D operations

  Tools for 3D operations with respect to 3D GIS are still lacking. Some examples of these tools are:
  - interactive 3D editing with realistic visualization
  - 3D overlay
  - implementation of point-in-tetrahedron testing
  - conversion between a 3D irregular network and 3D regular grid useful for many operations
  - a virtual reality interface for conveniently exploring content of a 3D database.

- Implementation of 3D constrained network construction

  Although the concept of constrained network construction using the raster approach has been generalized for n-dimensions, only the constrained triangulation was implemented in ISNAP. The constrained tetrahedronization still needs to be implemented.

- Development of 3D spatial index

  Many operations in 3D GIS, for example for realistic visualization, require data to be organized in a specific structure for efficiency. These are task-oriented index structures. The integrated database can only provide data in a basic structure and so requires the index structure appropriate for each task to be built on top of it. There are still requirements to identify index structure that provide efficient operation for each task. An object-oriented approach is potential for this kind of development; however, further studies, implementation and experiments are still needed. Such a study should include how to incorporate various database views and spatial index structures with the core database.

- Further tests for applicability of the simplicial network data model with some evaluations, for example against handling complex queries, finite element analysis, visualization.
- Methods to handle uncertainty covering 3D cases, for example to resolving lines or planes that coincide in 3D space,
- Maintenance of a 3D simplicial network spatial model; this is the problem of updating the spatial database which also requires the development of consistency rules.
- Comparative study raster and vector approaches for constrained network construction with respect to speed, memory and storage usage and overall efficiency.
- High quality 3D cartographic presentation of 3D spatial model, including:
  - design of 3D symbols
  - design of artificial texture
  - text and name placement in 3D space
  - use of 3D database for pictorial maps
  - 3D graphic generalization.

As described in the previous sections, this research work has covered a number of aspects of 2D and 3D spatial data structuring, data modelling, database populating, application development and user interface for a GIS. An important component of this research is the software that can be used for the development of an operational GIS system. Key aspects and problems were identified and have been implemented and tested. However, there are still other related issues that need to be investigated further and considered for future development. We also recommend the following further explorations:

- Implementing the constrained 3D TIN; incorporating 3D constrained features will extend data handling and provide more spatial information.
- Further developing and formalising 3D spatial data. This is an important task for describing the relationships and links between data and is the next logical step in the modelling process.

- Redesigning the OO spatial data model using object-oriented data analysis and design tools such as UML (United Modelling Language) to accommodate more complex situations.
- Developing and implementing spatial operators for spatial database manipulations.
- Constructing an advanced graphics user interface as the front-end program. Simple display interfaces have been developed. Although they were able to perform the task further work needs to be considered.
- Investigating an optimal integration between the developed subsystems with an object-oriented database management system (OO DMBS) as a database engine for facilitating TIN-based GIS.

Current research and industrial efforts for solving 3D spatial data information system appears to focus mainly on the non object-oriented approach, that is to say procedural (structured) techniques and the relational database. This book also explained how object-oriented techniques could be used for developing a TIN-based spatial data information system. The proposed subsystems work as described in chapter 8 and 9. The subsystems represent several of the major components that any GIS would have, that is from data input to a display or visualization subsystem. Although each of the proposed subsystems work and generate good results, they do not operate together as one fully operational system. Aspects of computing such as system integration, graphical user interface need to be further considered for full integration to be materialised.

All the developed algorithms have shown to work and provide a good framework for 3D GIS development. The performance of the algorithms in terms of computing yard sticks such as speed and data volume, however, is not part of the work discussed here.

Emerging technologies like Internet and Web influence the way we carry out research and certainly will play a major role in the next generation of 3D GIS aspects like data modelling, processing, databasing, analysis, and data distribution as discussed in the previous chapter.

Finally, we hope that all discussions from this book would lead to more interesting, advanced research output, and eventually a "true" 3D GIS could be realized in the near future.

References and Bibliography

Abdul-Rahman, A (1992) Triangular irregular network in digital terrain relief modelling. M. Sc. Thesis, ITC, Enschede, The Netherlands, 80 p.

Abdul-Rahman, A (2000) The design and implementation of a two and three-dimensional triangular irregular network based GIS. PhD Thesis, University of Glasgow, U.K., 204 pp.

Alagic, S (1989) Object-oriented database programming. Springer Verlag, New York, 320 pp.

Alexandroff, P (1961) Elementary concepts of topology. Dover Publications, Inc., New York.

Alia, A., Williams, H (1994) Approaches to the representation of qualitative spatial relationships for geographic databases. In: Molenaar, M., and De Hoop, S., (Eds.) Advanced geographic data modelling. Netherlands Geodetic commission, pp. 204-216

Anton, H (1987) Elementary linear algebra, Fifth edition, John Wiley & Sons.

Arc/Info (1991) Surface modelling with TIN. Arc/Info user's guide. ESRI, U.S.A.

Argiro, V. , Van Zandt, W (1992) Voxels: data in 3D, Byte, Vol. 17, May, pp. 177-182

Armstrong, M.A (1983) Basic topology, Springer, New York.

Avis, D., Bhattacharya, K.B (1983) Algorithms for computing d-dimensional Voronoi diagrams and their duals. Advances in Computing Research, 1, pp. 159-180

Ayugi, S.W.O (1992) The multi-valued vector map. M.Sc. Thesis, ITC, Enschede, The Netherlands

Bak, P.R.G, Mill, A.J.B (1989) Three dimensional representation in a geoscientific resource management system for minerals industry. In: Raper, J.(Ed.) Three dimensional applications in geographical information systems. Taylor & Francis, London, pp. 155-182

Barbalata, J.C., Lebel, R (1992) Digital elevation model for photogrammetric measurements of soil erosion. International Archives of Photogrammetry and Remote Sensing. Vol. XXIX, Part B4, Commission IV, Washington, D.C., U.S.A., pp. 831-835

Batten, L.G (1989) National capital urban planning project: development of a 3-D GIS. Proc. of GIS/LIS '89. ACSM/ASPRS. Falls Church, pp. 781-786.

Beaty, J.C., Booth, K.S (1982) Tutorial: computer graphics. Second Edition, IEEE Computer Society Press, Silver Spring, MD

Bernal, J(1988) On constructing Delaunay triangulation for a set of constrained line segments. Technical Note 1252, National Institute of Standards and Technology, United States of Commerce

Blum, H (1967) A transformation for extracting new descriptors of shape. Proceedings of Symposium on Models for Perception of Speech and Visual Form. MIT Press, Cambridge, Mass., pp. 362-380

Bonham-Carter, G. F (1996) Geographic information systems for geoscientists: modelling with GIS. Computer Methods in the Geosciences. Vol. 13, Pergamon Publications. 398 p

Booch, G (1994) Object-oriented analysis and design with applications, 2nd. Edition, Addison-Wesley Publishing Co., Menlo Park, CA., 589 p

Borgefors, G (1984) Distance transformations in arbitrary dimensions. Computer Vision, Graphics, and Image Processing. 27, pp. 321-345

Borgefors, G (1986) Distance transformations in digital images. Computer Vision, Graphics, and Image Processing. 34, pp. 344-371

Bouloucos, T. Ayugi, S.W.O, Kufoniyi, O (1993) Data structure for multi-valued vector maps, Proc. Fourth European Conference on Geographical Information Systems (EGIS'93). Genoa, Italy, pp. 237-245

Bouloucos. T, Kufuniyi, O., Molenaar, M (1990) A relational data structure for single valued vector maps. International Archives of Photogrammetry and Remote Sensing, Vol. 28, Part 3/2, Commission III, Wuhan, China, pp. 64-74

Bowyer, A (1981) Computing Dirichlet tessellation. Computer Journal, 24, pp. 162-166

Brassel, K.E, Reif, D (1979) Procedure to generate Thiessen polygons. Geographical Analysis. 11, pp. 289-303

Bric, V (1993) 3D vector data structures and modelling of simple objects in GIS. M. Sc. Thesis, ITC, Enschede, The Netherlands, 107 p

Bric, V, Pilouk, M, Tempfli, K (1994), Towards 3D-GIS: Experimenting with a Vector Data Structure. Proc. of the Symposium on Mapping and Geographic Information Systems. Georgia, USA, ISPRS Archives Vol. 30, Part 4, pp. 634-640

Bric, V, Pilouk, M (1994) Computation of topologic space. ITC, Enschede, The Netherlands

Bric, V, Pilouk, M, Tempfli, K (1994). Towards 3D-GIS: Experimenting with a vector data structure. International Archives of Photogrammetry and Remote Sensing. Vol. XXX, Part 4, Athens, Georgia, USA, pp. 634-640.

Bric, V (1993) 3D vector data structures and modelling of simple objects in GIS. M.Sc. Thesis, ITC, Enschede, The Netherlands.

Brockschmidt, K (1993) Programming for Windows with Object Linking and Embedding 2.0. Microsoft Press

Brunet, P (1992) 3-D structures for the encoding of geometry and internal properties, In: Three-Dimensional Modeling with Geosciencetific Information Systems by A. K. Turner (ed.). NATO ASI Series C, Kluwer Academic Publishing, Dordrecht, Vol. 354, pp. 159-188

Burrough, P.A (1986) Principles of geographical information systems for land resources assessment. Clarendon Press, Oxford University Press, 194 pp

Cambray, B. de, (1993) Three-dimensional (3D) modelling in a geographical database. Proc. 11th International Symposium on Computer Assisted Cartography (AUTOCARTO 11). Minneapolis, pp. 338-347.

Cambray, de B, Yeh, T. S (1994) A multidimensional (2D, 2.5D, and 3D) geographical data model. International Conference on Management of Data

(COMAD'94). Bangalore, India, Tata Mc Graw-Hill, pp.317-336, http://www.prism.uvsq.fr/public/beatrix/publi_en.html

Cantor, G (1880) Über unendliche, lineare Punktmannigfaltigkeiten. Math. Ann. (B) 17, pp. 355-388

Carlson, E (1987) Three dimensional conceptual modeling of subsurface structures. Technical Papers, Vol. 4, ASPRS-ACSM Annual Convention. Baltimore, Maryland, pp. 188-200

Chen, P. PS (1983) The entity-relationship approach to information modelling & analysis. Proc. International Conference. North-Holland

Chen, TK, and Abdul-Rahman, A (2006) 0D feature in 3D planar polygon testing for 3D spatial analysis. In: Abdul-Rahman, Coors, Zlatanova (eds.) Innovations in 3D geo information systems. Springer. Germany

Chen, X., Ikeda, K., Yamakita, K., Nasu, M (1994) Raster algorithms for generating Delaunay tetrahedral tessellations. International Archives of Photogrammetry and Remote Sensing. Commission III, Vol. 30, Part 3/1, Munich, Germany, pp. 124-131

Chew, LP (1989). Constrained Delaunay triangulations. Algorithmica 4, pp. 97-108

Chhatkuli, RR (1993) Modelling data quality parameters in a multiple-theme vector data structure and its implementation in a geographic information system. M.Sc. Thesis, ITC, Enschede, The Netherlands

Collin, WJ (1992) Data structures: an object-oriented approach. Addison-Wesley. Reading Massachusetts, 624 pp

Cöltekin A (2002) An Analysis of VRML-based 3D Interfaces for Online GIS: Current Limitations and Solutions. Surveying Science in Finland. Vol.20, No: 1-2, p.80-91

Connolly, T, Begg, C (2002) Database Systems – A Practical Approach To Design, Implementation and Management. 3rd Edition, Pearson Addison-Wesley. Menlo Park, California.

Coors V (2003) 3D-GIS in Networking Environments. Computer, Environments and Urban Systems, Vol. 27/4, 2003, Special Issue 3D cadastre, pp 345-357.

Coors V, Jung, V (1998) Using VRML as an Interface to the 3D Data Warehouse. Proceedings of the third symposium on the Virtual reality modeling language, Monterey, California, United States , pp 121-140.

Corbett, JP (1979) Topological principles in cartography. Technical paper 48, U.S. Department of Commerce, Bureau of the census, 50 pp

Dahl, OJ, Myrhaug, B, Nygaard, K (1970) SIMULA common base language. Norwegian Computing Center S-22, Oslo, Norway

Danielsson, P.E (1980) Euclidean distance mapping. Computer Graphics and Image Processing. 14, pp. 227-248

Date, C.J (1986) An introduction to database systems. Vol. 1, Addison-Wesley, Reading, Mass.

De Floriani, L, Puppo, E (1988) Constrained Delaunay triangulation for multi resolution surface description. Proc. of the 9th International Conference on Pattern Recognition. Rome, Italy, pp. 566-569

De Floriani, L, Puppo, E (1992) An on-line algorithm for constrained Delaunay triangulation. CVGIP: Graphical Models and Image Processing. 54, pp. 290-300

Delaunay, B (1934) Sur la sphére vide. Bulletin of the Academy of Sciences of the USSR. Classe des Sciences Mathématiques et Naturelles, 8, pp. 793-800

Delobel, C., Lecluse, C, Richard, P (1995) Database: from relational to object-oriented systems. International Thomson Computer Press. London, 382 p

Devlin, K (1994) Mathematics: The sciences of patterns. Scientific American Library, New York, 215 pp

Diehl S. (2001) Distributed Virtual Worlds. Springer-Verlag, Berlin Heidelberg New York

Dirichlet, G.L (1850) Über die reduction der positiven quadratischen formen mit drei unbestimmten ganzen zalen, J. Reine u. Angew. Math. 40, pp. 209-227

DLG-E, *Digital Line Graph-Enhanced*, U.S. Department of the Interior, U.S. Geological Survey

Dong, F (1996) Three-dimensional models and applications in subsurface modeling. Department of Geomatics Engineering Reports No. 20093. University of Calgary, 93 p

Dwyer, R.A (1987) A fast divide-and-conquer algorithm for constructing Delaunay triangulations. Algorithmica. Vol. 2, pp. 137-151

Dykes J.A, Moore K.E, Fairbairn D (1999) From Chernoff to Imhof and Beyond: VRML and Cartography. Proc. of 4th Int. Conference on the VRML and Web3D Technologies (VRML99). Paderborn, Germany.

Ebner, H, Eder, K (1992) State-of-the-art in digital terrain modelling. Proc. 3rd. European Conference on Geographical Information Systems (EGIS'92). Volume. 1, Munich, Germany, pp. 681-690.

Ebner, H, Hossler, R, Wurlander, R (1990) Integration of an efficient DTM program package into geographical information systems. International Archives of Photogrammetry and Remote Sensing. Vol. 28, Part 4, Commission IV, Tsukuba, Japan, pp. 116-121

Edelsbrunner, H, Preparata, FP, West, DB (1986) Tetrahedrizing point sets in three dimensions. Technical Report UIUCDCS-R-86-1310. Department of Computer Science, University of Illinois, 1304 W. Springfield Avenue, URBANA, Il 61801

Egenhofer, MJ (1991) Extending SQL for cartographic display. Cartography and Geographic Information Systems. Vol. 18, No. 4, pp. 230-245

Egenhofer, MJ (1990) Interaction with geographic information systems via spatial queries. Journal of Visual Languages and Computing. Vol. 1, No. 4, pp. 389-413

Egenhofer, MJ (1989) A formal definition of binary topological relationships. Technical Report No. 101. NCGIA/Department of Surveying Engineering, University of Maine. Orono, ME, USA

Egenhofer, MJ, Frank, AU (1989) Object-oriented modelling in GIS: Inheritance and propagation. Proc. 9th International Symposium on Computer Assisted Cartography (AUTOCARTO 9). Baltimore, Maryland, pp. 588-589

Egenhofer, MJ, Frank, AU, Jackson, JP (1989) A topological data model for spatial databases. NCGIA Technical Report, No. 104

Egenhofer, MJ, Franzosa, RD (1991) Point-set topological spatial relations. International Journal for Geographical Information Systems. Vol. 5, No. 2, pp. 161-174

Egenhofer, MJ, Herring, JR (1990) A mathematical framework for the definition of topological relationships. Proc. of the Fourth International Symposium on Spatial Data Handling. Zurich, Switzerland, pp. 803-813.

Egenhofer, MJ, Herring, JR (1992) Categorizing binary relationships between regions, lines, and points in geographical databases. Technical report, Department of Surveying Engineering, University of Main, USA

Ehlers, M, Greenlee, D, Smith, T, Star, J (1991) Integration of remote sensing and GIS: Data and data access. Photogrammetric Engineering & Remote Sensing, Vol. 57, No. 6, pp. 669-675

Ehlers, M, Edwards, G, Bédard (1989) Integration of remote sensing with geographic information systems: a necessary evolution. Photogrammetric Engineering & Remote Sensing. Vol. 55, No. 11, pp. 1619-1627

Fang, T P, Piegl, LA (1993) Delaunay triangulation using a uniform grid. IEEE Computer Graphics & Applications. May 1993, pp. 36-47

Fang, T P, Piegl, LA (1995) Delaunay triangulation in three dimensions, IEEE computer Graphics & Applications. September, 1995, pp. 62-69

Field, AD (1986) Implementing Watson's algorithm in three dimensions. Proc. of ACM Symposium on Computational Geometry. pp. 246-259

Field, AD, Smith, WD (1991) Graded tetrahedral finite element meshes. International Journal for Numerical Methods in Engineering. 31(3), pp. 413-425

Finkbeiner, DT, Lindstrom, WD (1987) A primer of discrete mathematics. W.H. Freeman and Company. New York, 363 pp

Fisher, TR (1993) Use of 3D geographic information systems in hazardous waste site investigations. In: Goodchild, M.F., Parks, B., and Steyaert, L., (Eds.), Environmental Modeling with GIS. Oxford University Press. New York

Flankin, WM (1984) Cartographic errors symptomic of underlying algebra problems. In: Marble, D, et al. (Eds.) Proc. of the International Symposium on Spatial Data Handling. Zurich, Switzerland

Flavin, M (1981) Fundamental concepts of information modelling. Yourdon Press Computing Series. Prentice-Hall, Inc., Englewood Cliffs, New Jersey, USA 128 pp

Flowerdew, R (1991) Spatial data integration. In: Maguire, DJ, Goodchild, MF, Rhind, DW, (Eds.) Geographical information systems principles and applications. Longman Scientific & Technical, pp. 375-387

Foley, JD, van Dam, A, Feiner, SK, Hughes, JF (1992) Computer graphics: principles and practice. Second Edition, Addison-Wesley. USA, 1175 pp

Förstner, W, Pallaske, R (1993) Mustererkennung und 3D-Geoinformationssysteme. ZPF, 61. Jg., 5/1993, pp. 167-177

Forstner, W (1995) GIS - the third dimension, Workshop on Current Status and Challenges of Geoinformation Systems. IUSM working group on LIS/GIS. University of Hannover, Germany, pp. 65-72

Frank, AU (1992) Spatial concepts, geometric data models and data structures. Computer & Geosciences. Vol. 18, No. 4, pp. 409-417

Frank, AU, Kuhn, W (1986) Cell graphs: a provable correct method for the storage of geometry. Proc. of the Second International Symposium on Spatial Data Handling. Seattle, Washington, USA, pp. 411-436

Fréchet, M (1906) Sur quelques points du calcul fonctionnel. Rendiconti di Palermo 22, 1-74

Fritsch, D (1990) Towards three-dimensional data structures in geographic information systems. Proc. First European Conference on Geographical Information Systems (EGIS'90). Volume 1, Amsterdam, The Netherlands, pp. 335-345

Fritsch, D, Pfannenstein, A (1992a) Integration DTM data structures into GIS data models. International Archives of Photogrammetry and Remote Sensing. Vol. XXIX, Part B3, Commission III, Washington, D.C., USA., pp. 497-503

Fritsch, D, and Pfannenstein, A (1992b) Conceptual models for efficient DTM integration into GIS. Proc. Third European Conference on Geographical Information Systems (EGIS'92). Volume. 1, Munich, Germany, pp. 701-710

Fritsch, D, Schmidt, D (1995) The object-oriented DTM in GIS. Proc. of Photogrammetric Week. Stuttgart, pp. 29-34

Fritsch, D (1996) Three-dimensional geographic information systems - status and prospects. International Archives of Photogrammetry and Remote Sensing (ISPRS). Vienna, Vol. 31, Part 4, pp. 215-221

Gargantini, I (1992) Modelling natural objects via octrees. In: Three-dimensional modeling with geoscientific by A. K. Turner (Ed.). NATO ASI Series, Kluwer Academic Publishers. pp. 145-157

Gatrell, AC (1991) Concepts of space and geographical data. In: Maguire, DJ, Goodchild, MF, Rhind, DW (Eds.). Geographical Information Systems. Vol. 1: Principles. Longman. UK

Geographical Information Science, Vol. 17, No.5, pp. 411-430

GEONOVA (2003) *Newsletter Q3/2003*

Giblin, P (1977) Graphs, surfaces and homology. Chapman and Hall, London

Gröger, G, Kolbe, T, Dress, R, Müller, H, Knopse, F, Gruber, U, and Krause, U (2004) Das interoperable 3D-Stadtmodell der SIG 3D der GDI NRW. Version 2. Stand: 10.5.2004
(URL: http://www.ikg.uni-bonn.de/sig3d/docs/Handout_04_05_10.pdf)

Goldberg, A, Robson, D (1983) Smalltalk-80: the language and its implementation. Addison-Wesley, Reading, Massachusetts

Golda, YV (1992) The "flowing" accumulation method and its application for earth surface analysis. International Archives of Photogrammetry and Remote Sensing. Vol. XXIX, Part B4, Commission IV, Washington, D.C., USA. pp. 836-842

Gorte, B, Koolhoven, W (1990) Interpolation between isolines based on the Borgefors distance transform. ITC Journal - Special Issue Remote Sensing and GIS, 1990-3. pp. 245-247

Green, P J, and Sibson, R (1978) Computing Dirichlet tessellations in the plane. Computer Journal, 21, pp. 168-173

Gruen, A, Streilein, A, Stallmann, D, Dan, H (1993) Automation of house extraction from aerial and terrestrial images. Conference ASIA. Wuhan, China

Guptill, C, Morrison, JL (1995) Elements of spatial data quality. Elsevier Science Ltd.

Guo, W (1996) Three-dimensional representation of spatial object and topological relationships. International Archives of Photogrammetry and Remote Sensing. Vol. XXXI, Part B3, Commission 3, K. Kraus and P. Waldhausl (eds.), XXXI International Congress of Photogrammetry and Remote Sensing, Vienna, pp. 273-278

Gütting, R (1988) Geo-relational algebra: a model and query language for geometric database system. In: Schmidt, J, Ceri, S, Missikoff, M (Eds.) Advances in Database Technology--EDBT '88, International Conference on Extending Database Technology. Venice, Italy, Lecture Notes in Computer Science, Springer Verlag, New York, Vol. 303, pp. 506-527

Güting, H (1994) An Introduction to Spatial Database Systems. VLDB Journal 3, Hans-J. Schenk (Ed.), pp.357-399, 1994

Guttman, A (1984) A dynamic index structure for spatial searching. Proc.of the SIGMOD Conference. Boston, pp. 47-57

Hausdorff, F (1914) Grundzüge der mengenlehre. Leipzig. Reprinted by Chelsea, New York. 88 pp

Hawryszkiewycz, IT (1991) Introduction to system analysis and design. Second edition, Prentice Hall, Australia

Hearnshaw, HM, Unwin, DJ (1994) Visualization in geographical information systems. Wiley and Sons, 243 pp

Herring, JR (1989) A fully integrated geographic information system. Proc. 9th International Symposium on Computer Assisted Cartography (AUTOCARTO 9). pp. 828-837

Herring, RJ, Egenhofer, MJ (1990) A mathematical framework for the definition of topological relationships. Proc. of the Forth International Symposium on Spatial Data Handling. Zurich, Switzerland, pp. 803-813

Herring, J, Larsen, R, Shivakumar, J (1988) Extensions to the SQL language to support spatial analysis in a topological data base. Proc. GIS/LIS '88. San Antonio, Texas, pp. 741-750

Hesse, W, Leahy, FJ (1990) Authoritative topographic-cartographic information system (ATKIS). The Project of the State Survey Authorities for the Creation of Digital Landscape Models and Digital Cartographic Models. Landesvermessungamt Nordrhein-Westfalen, Bonn, 29 pp

Houlding, S W (1994) 3D geoscience modelling: computer techniques for geological characterization. Springer-Verlag, Berlin, 309 p

Howe, DR (1989) Data analysis for database design. Second edition, Edward Arnold A Division of Hodder & Stoughton, London. 317 pp

Illustra, (1994) Illustra, relational databases and spatial data. An Illustra Technical White Paper by Malcolm Colton, Oakland, CA, 6 pp

Institute for Photogrammetry (ifp) (1997) Working Group IV/2: Digital Terrain Models, Orthoimages, and 3D GIS. University of Stuttgart, Germany. http://www.ifp.uni-stuttgart.de/comm4/wgIV2.html

Intergraph, (1995) New OLE extensions for CAD/CAM/CAE and GIS adopted, Press Releases. Intergraph Corp., Huntsville, http://www.intergraph.com/press95/dmpr.html.

Jackins, C L, Tanimoto, SL (1980) Oct-trees and their use in representing three-dimensional objects. Computer Graphics and Image Processing. Vol. 14, pp. 249-270

Jackson, J (1989) Algorithms for triangular irregular networks based on simplicial complex theory. ASPRS-ACSM Annual Convention. Baltimore, MD, USA., pp. 131-136

Jackson, MA (1983) *System development*, Prentice Hall, 418 pp.

Jansen R (2003) Oracle, Java, XML: Integration in Oracle9i. Frankfurt, Germany.

Jianya, G, Deren, L (1992) Object-oriented data models in GIS. International Archives of Photogrammetry and Remote Sensing. Vol. XXIX, Part B3, Commission 3, W. Fritz and J. R. Lucas (eds.), XXIX International Congress of Photogrammetry and Remote Sensing, Washington, pp. 773-779

Joe, B (1989) Three-dimensional triangulations from local transformations. Siam Journal on Scientific and Statistical Computing, 10(4), pp. 718-741

Jones, CB (1989) Data structures for three-dimensional spatial information systems in geology, International Journal of Geographical Information Systems. Vol.3, No. 1, Taylor & Francis, London. pp. 15-31

Kainz, W (1989) Order, topology and metric in GIS. ASPRS-ACSM Annual Convention, Baltimore, Vol. 4, pp. 154-160

Kainz, W (1990) Spatial relationships--topology versus order. Proc.of the Fourth International Symposium on Spatial Data Handling. Zurich, Switzerland, Brassel, K, and Kishimoto, H, (Eds.), Vol. 2, pp. 814-819

Kainz, W, Egenhofer, M, Greasley, I (1993) Modeling spatial relations and operations with partially ordered sets. International Journal of Geographical Information Systems, Vol. 7, No. 3, pp. 215-229

Kainz, W, Shahriari, N (1993) Object-oriented tools for designing topographic databases. Proc. GIS/LIS'93. pp. 341-350

Kanaganathan, S, Goldstein, NB (1991) Comparison of four-point adding algorithms for Delaunay-type three-dimensional mesh generators. IEEE Transactions on Magnetics. 27(3), pp. 3444-3451

Kavouras, M, Masry, S (1987) An information system for geosciences, design considerations. Proc. 8th International Symposium on Computer Assisted Cartography (AUTOCARTO 8). Baltimore, MD, pp. 336-345

Kemp, Z (1990) An object-oriented model for spatial data. Proc. 4th. International Symposium on Spatial Data Handling. Vol. 2, Zurich, Switzerland, pp. 659-668

Kinsey, LC (1993) Topology of surfaces. Springer Verlag. New York, 276 pp

Knuth, DE (1973) The Art of computer programming. Vol. 3: Sorting and Searching, Addison-Wesley, Reading

Kofler, M (1998) R-trees for Visualizing and Organizing Large 3D GIS Databases. Dissertation TU Graz, Austria

Kolbe, A (2003) Applications and Solutions for Interoperable 3D Geovisualization. Proc. of the Photogrammetric Week 2003. Stuttgart, Germany.

Kraak, MJ (1992) Working with triangulation-based spatial data in 3D space. ITC Journal 1992-1, pp. 20-33

Kraak, MJ, Verbree, E (1992) Tetrahedrons and animate maps in 2D and 3D space. In: Proc. of the 5th International Symposium on Spatial Data Handling. pp. 63-71

Kraus, K (1995) From digital elevation model to topographic information system. 45th.Photogrammetric Week. D. Fritsch and D. Hubbie (eds.), Stuttgart, pp. 277-285

Kufoniyi, O (1989) Editing of topologically structured data. M.Sc. Thesis. ITC, Enschede, The Netherlands

Kufoniyi, O, Bouloucos, T (1994) Flexible integration of terrain objects and DTM in vector GIS. Proceedings International Colloquium on Integration, Automation and Intelligence in Photogrammetry, Remote Sensing and GIS. Wuhan, pp. 111-122

Kufoniyi, O (1995) Spatial coincidence modelling, automated database updating and data consistency in vector GIS. Ph.D. Thesis. Wageningen Agricultural University, The Netherlands, 206 pp

Kufoniyi, O (1995b) An introduction to object-oriented data structures. ITC Journal 1995-1, pp. 1-7

Kuhn, W, Frank, AU (1991) A formalization of metaphors and image-schemas in user interfaces. In: Mark, D, Frank, A (Eds.) Cognitive and linguistic aspects of geographic space. Kluwer Academic Publ., Dordrecht, pp. 419-434

Kwan, MP, Lee, J (2003) Emergency response after 9/11: the potential of real-time 3D GIS for quick emergency response in micro-spatial environments. (http://dx.doi.org/10.1016/j.compenvurbsys.2003.08.002).

Lammersen R. van, Hoogerwerf ,T (2003) Geo-Virtual reality and Participatory Planning. CGI Report 2003-07, Wageningen, The Netherlands

Langran, G (1992) Time in geographic information systems. Taylor & Francis. London.

Larkin, BJ (1991) An ANSI C program to determine in expected linear time the vertices of the convex hull of a set of planar points. Computers & Geosciences. 17, pp. 431-443

Lattuada, R, Raper, J (1995) Applications of 3D Delaunay triangulation algorithms in geoscientific modeling. GISRUK'95 Conference. U.K, http://www.bbk.ac.uk/department/ geography/jonathanraper.html

Laurini, R, Thompson, D (1993) Fundamentals of spatial information systems. *Academic Press*, London, 680 p

Laurini, R (2001) Information System For Urban Planning – A hypermedia cooperative approach. London New York

Lawson, CL (1985) Some properties of n-dimensional triangulation. External Report, JPL Publication 85-42. National Aeronautics and Space Administration.

Lawson, CL (1977) Software for C1 surface interpolation. In Rice, J (Ed.) Mathematcal Software III. Academic Press. Newyork, USA, pp. 161-194

Lawson, CL (1972) Generation of a triangular grid with application to contour plotting. California Institute of Technology, Jet Pollution Laboratory. Technical Memorandum No. 299

Leach, R (1995) Object-oriented design and programming with C++. AP Professional. London. 463 p

Lee, DT, Lin, AK (1986) Generalized Delaunay triangulation for planar graph. Discrete & Computational Geometry. 1, pp. 201-217

Lee, DT, Schachter, BJ (1980) Two algorithms for constructing a Delaunay triangulation. International Journal of Computer and Information Sciences. 9, pp. 219-242

Lopez, X (2003) Oracle Database 10g: A Spatial VLDB Case Study. Oracle Cooperation Whitepaper. (URL:http://otn.oracle.com/products/spatial/pdf/customer_success/papers/spatial_10g_ow2003.pdf).

Lewis, BA, Robinson, JS (1978) Triangulation of planar regions with applications. Computer Journal. 21, pp. 324-332

Lingas, A (1986) The greedy and Delaunay triangulations are not bad in the average case. Information Processing Letters. 22, pp. 25-31

Li, R (1993) Three-dimensional GIS: a simple extension in the third dimension? ACSM/ASPRS Annual Convention. New Orleans, USA. Vol. 3, pp. 218-227

Li, R (1994) Data structures and application issues in 3-D geographic information systems. Geomatica. Vol. 48, No. 3, pp. 209-224

Li, R, Chen, Y, Dong, F, Qian, L, Hughes, JD (1996) 3D data structures and applications in geological subsurface modelling. International Archives of Photogrammetry and Remote Sensing. Vol. XXXI, Part B4, Commission 4, K. Kraus and P. Waldhausl (eds.), XXXI International Congress of Photogrammetry and Remote Sensing, Vienna, pp. 508-513

Liu, CL (1986) Elements of discrete mathematics. Second edition, McGraw-Hill, 433 pp

Nebiker, S (2003) Support For Visualization and Animation in a Scalable 3D GIS Environment: Motivation, Concepts and Implementation. International Archives of Photogrammetry, Remote Sensing and Spatial Information Science, Vol. XXXIV-5/W10

Macedonio, G, Pareschi, MT (1991) An algorithm for the triangulation of arbitrarily distributed points: applications to volume estimate and terrain fitting. Computer & Geosciences. 17, pp. 859-874

Maguire, DJ, Dangermond, J (1991) Functionality of GISs. In: Maguire, DJ, Goodchild, MF, Rhind, DW (Eds.) Geographical Information Systems. Vol. 1, Principles, Harlow: Longman Scientific & Technical, pp. 319-335

Maguire, D.J., Goodchild, M.F., and Rhind, D.W., (Eds), 1991, Geographical information systems: principles and applications. Longman Scientific & Technical.

Makarovic, B (1984) Structures for geo-information and their application in selective sampling for digital terrain models. ITC Journal 1984-4, pp. 285-295

Makarovic, B, (1977) Composite sampling for DTMs, ITC Journal

Males, RM (1978) ADAPT - a spatial data structure for use with planning and design models. In: Dutton, G., (Ed.), First International Symposium on Advance Study on Topological Data Structures for Geographic Information Systems, Vol. 3, 19 pp

Manacher, GK, Zobrist, AL (1979), Neither the greedy nor the Delaunay triangulation of a planar point set approximates the optimal triangulation. Information Processing Letters, 9, pp. 31-34

Mäntylä, M (1988) Solid modelling. Computer Science Press. Rocville, Maryland, 401 pp

MAP24 (2004). (http://www.map24.de/).

Marble, DF, Calkins, HW, Peuquet, DJ (1984) Technical description of the DIME system. Basic Readings in Geographic Information Systems. SPAD Systems, Ltd. USA., pp. 57-64

Mark, DM, Cebrian, JA (1986) Octrees: a useful data-structure for the processing of topographic and sub-surface data. Technical Papers of ACSM-ASPRS Annual Convention. Vol. 1 (Cartography and Education)

Mark, DM, Lauzon, JP, Cebrian, JA (1989) A review of quadtree-based strategies for interfacing coverage data with digital elevation models in grid form. International Journal of Geographical Information Systems. Vol.3, No. 1, Taylor & Francis, London, pp. 3-14

Martin, J (1983) Managing the data-base environment. Prentice-Hall, Inc., Englewood Cliffs, New Jersey

Maus, A (1984) Delaunay triangulation and the convex hull of n points in expected linear time. BIT, 24, pp. 151-163

McCullagh, MJ, Ross, CG (1980) Delaunay triangulation of a random data set for isarithmatic mapping. The Cartographic Journal. 17, pp. 93-99

Meagher, D (1982) Geometric modelling using octree encoding. Computer Graphics and Image Processing. Vol. 19, pp. 129-147

Meier, A (1986) Applying relational database techniques to solid modelling. CAD. Vol.18, No.6, pp. 319-326

Meij, L. v.d (1992) Topologische relaties en bevragingen in de formele datastructuur voor drie-dimensionele vectorkaarten. Scriptie, LU Wageningen, The Netherlands

Microsoft, (1993) Object linking and embedding: OLE 2.0 design specification, Microsoft Corporation

Midtbø, T (1996) Spatial modelling by delaunay networks of two and three dimensions. PhD thesis. Norwegian Institute of Technology. University of Tronheim, Norway, http://guran1.iko.unit.no/home/terjem/terjem.html

Midtbø, T (1993a) Incremental Delaunay tetrahedrization for adaptive data modelling. Proc. Fourth European Conference on Geographical Information Systems (EGIS'93). Genoa, Italy, pp. 227-236

Midtbø, T (1993b) Spatial modelling by Delaunay networks of two and three dimensions, Dr. Ing. Thesis, Norwegian Institute of Technology, University of Tronheim, Norway, 147 pp

Miller, CL, Laflamme, RA (1958) The digital terrain model - theory and application. Photogrammetric Engineering. pp. 433-442

Mirante, A, Weingarten, N (1982) The radial sweep algorithm for constructing trianguled irregular networks. IEEE Computer Graphics and Applications. 2, pp. 11-21

Moellering, H (1991) Spatial database transfer standards: current international status. Elsevier Applied Science. 247 pp

Moise, EE (1977) Geometric topology in dimension 2 and 3 Springer Verlag, New York

Molenaar, M, Fritsch, D, Bill, R (1996) Conceptual aspects of GIS technology. ISPRS Congress Tutorial, Vienna

Molenaar, M (1994a) A syntax for representation of fuzzy spatial objects. In: Molenaar, M, De Hoop, S (Eds.) Advanced geographic data modelling: spatial data modelling and query languages for 2D and 3D applications. Netherlands Geodetic Commission, No. 40, Delft, The Netherlands, pp. 155-169.

Molenaar, M (1994b) A syntactic approach for handling the semantics of fuzzy spatial objects. European Science Foundation, GISDATA, Baden, Austria, 15 pp

Molenaar, M (1993) Object hierarchies and uncertainty in GIS or why is standardisation so difficult?, Geo-Informations-Systeme. Vol. 6, No. 4, pp. 22-28

Molenaar, M (1992) A topology for 3D vector maps. ITC Journal. 1992-1, pp. 25-33

Molenaar, M (1991) Formal data structures, object dynamics and consistency rules. Digital Photogrammetric Systems. Herbert Wichmann Verlag GmbH, Karlsruhe, pp. 262-273

Molenaar, M (1990) A Formal data structure for 3-D vector maps. Proceedings First European Conference on Geographical Information Systems (EGIS'90). Volume. 2, Amsterdam, The Netherlands, pp. 770-781.

Molenaar, M (1989) Single valued vector maps - a concept in GIS, Geo-Informations-Systeme. Vol. 2, No. 1, pp. 18-27

Molenaar, M (1988) Single valued polygon maps. International Archives of Photogrammetry and Remote Sensing. Vol. 27, Part B4, Commission IV, Kyoto, Japan, pp. 592-601

Ning, S (1992) On the principles and the approaches of implementing the strict digital geometric rectification for SPOT imagery. International Archives of Photogrammetry and Remote Sensing. Vol. XXIX, Part B3, Commission III, Washington, D.C., USA., pp. 32-34

OGC (2001) Web Map Service Implementation Specification. (http://www.opengis.org/docs/01-068r2.pdf)

OGC (2002) Web Feature Service Implementation Specification. (www.opengis.org/docs/02-058.pdf).

OGC (2003a) Geographic Markup Language (GML 3). (http://www.opengis.org/docs/02-023r4.pdf).

OGC (2003b) OpenGIS Reference Model. (http://www.opengis.org/docs/03-040.pdf).

Oosterom P van, Stoter J, Quak W, Zlatanova S (2002) The Balance Between Geometry and Topology. Proc. of Spatial Data Handling. Ottawa, Canada

Oosterom, P. van (1990) Reactive data structures for geographic information systems, PhD Thesis. Leiden University. The Netherlands, 197 pp

Orenstein, JA (1990) An object-oriented approach to spatial data processing. Proc. of the 4th International Symposium on Spatial Data Handling. Vol. 2, Zurich, Switzerland, pp. 669-698

Peng, W, Molenaar, M (1995) An object-oriented approach to automated generalization. Proc. of GeoInformatics '95. Vol. 1, Hong Kong, pp. 295-304

Peng, W, Tempfli, K, Molenaar, M (1995) Automated generalization in a GIS context. Proceedings of GeoInformatics '96. Florida, USA, 11 pp

Peng, Y.R, Tsou, MH (2003) Internet GIS – Distributed Geographic Information Services for the Internet and Wireless Networks. Hoboken, New Jersey, USA.

POET Software Corporation,(1996) Why use an ODBMS?. POET Technical References, http://www.poet.com/t_oovsre.htm#ODBMS

Petchenik, BB (1991) New directions for national mapping. URISA, Vol.3, No.1, pp.77-79

Petrie, G, Kennie, TJM (1990) Terrain modelling in surveying and civil engineering. Whittles Publishing. Glasgow, 351 p

Peucker, T, Chrisman, N (1975) Cartographic data structures. The American Cartographer. Vol. 2, No. 2, pp. 55-69

Peucker, T K (1978) Data structures for digital terrain models: discussion and comparison. 1st. International Advanced Study Symposium on Topological Data Structures for Geographical Information Systems. Harvard Paper on GIS, Edited by G. Dutton, Vol. 5

Peuquet, DJ (1988) Representations of geographic space: toward a conceptual synthesis. Annals of the Association of American Geographers. 78, pp. 375-94

Peuquet, DJ (1986) The use of spatial relationships to aid spatial database retrieval. Proc. of the Second International Symposium on Spatial Data Handling, Seattle, WA

Peuquet, DJ (1984) A conceptual framework and comparison of spatial data models. CARTOGRAPHICA. Vol. 21, No. 4, pp. 66-113

Pfannenstein, A, Reinhardt, W (1993) Data analysis in geographical information systems in combination with integrated digital terrain models. Proc. Fourth European Conference on Geographical Information Systems (EGIS'93). pp. 1341-1349

Pigot, S (1992) A topological model for a 3-D spatial information system. Proc. 5th International Symposium on Spatial Information Handling. Charleston, S.C., pp. 344-360

Pigot, S (1991) Topological models for 3D spatial information systems. Proc. 10th International Symposium on Computer Assisted Cartography (AUTOCARTO 10). Technical Papers. ACSM-ASPRS, Annual Convention, Vol. 6, Baltimore, Maryland, USA., pp. 369-391

Pilesjo, P, Michelson, DB, Hall-Konyves, KM (1992) Digital elevation models for identification of potential wetlands. International Archives of Photogrammetry and Remote Sensing. Vol. XXIX, Part B4, Commission IV, Washington, D.C., USA., pp. 817-822

Pilouk, M, Radjabi Fard, A, Tempfli, K (1994) Local updating of TIN for the integrated DTM and GIS data structure. International Archives of Photogrammetry and Remote Sensing. Vol. XXX, Part 4, Athens, Georgia, USA, pp. 460-466

Pilouk, M, Kufoniyi, O (1994) A relational data structure for integrated DTM and multitheme GIS. International Archives of Photogrammetry and Remote Sensing. Commission III, Vol. 30, Part 3/2, Munich, Germany, pp. 670-677

Pilouk, M, Tempfli, K, Molenaar, M (1994) A tetrahedron-based 3D vector data model for geoinformations. In: Molenaar, M, De Hoop, S (Eds.) Advanced geographic data modelling: spatial data modelling and query languages for 2D and 3D applications. Netherlands Geodetic Commission, No. 40, Delft, The Netherlands, pp. 129-140

Pilouk, M, Tempfli, K (1994) An object oriented approach to the unified data structure of DTM and GIS. International Archives of Photogrammetry and Remote Sensing. Vol. XXX, Part 4, Athens, Georgia, USA, pp. 672-679

Pilouk, M, Tempfli, K (1994) Integrating DTM and GIS using a relational data structure. GIS'94. Vol. 1, Vancouver, Canada, pp. 163-169

Pilouk, M, Tempfli, K (1993) An integrated DTM-GIS data structure: a relational approach. Proc. of 11th International Symposium on Computer Assisted Cartography (AUTOCARTO 11). Minneapolis, Minnesota, USA, pp. 278-287

Pilouk, M, Tempfli, K (1992) A digital image processing approach to creating DTMs from digitized contours. International Archives of Photogrammetry and Remote Sensing, Vol. XXIX, Part B4, Commission IV, Washington, D.C., USA., pp. 956-961

Pilouk, M (1992) Fidelity improvement of DTM from contours. M.Sc. Thesis. ITC, Enschede, The Netherlands

Pilouk, M (1996) Integrated Modelling for 3D GIS. PhD Thesis. ITC Publication No. 40, 200 p

Pohl, I (1993) Object-oriented programming using C++. Benjamin/Cummings Publishing Company, Inc., California, 496 pp

Pullar, D (1988) Data definition and operators on a spatial data model. ACSM-ASPRS, Annual convention. Vol. 2, pp. 196-202

Pullar, D, Egenhofer, M (1988) Towards formal definitions of topological relations among spatial objects. Proc. of the 3rd. International Symposium on Spatial Data Handling. Sydney, Australia, pp. 225-242

Qingquan, L, Deren, L (1996) Hybrid data structure based on octree and tetrahedron in 3-D GIS. International Archives of Photogrammetry and Remote Sensing. Vol. XXXI, Part B, Commission 4, K. Kraus and P. Waldhausl (eds.), International Congress of Photogrammetry and Remote Sensing, Vienna, pp. 503-507

Raper, J (1992) Key 3D modelling concepts for geoscientific analysis. In: Three-dimensional modeling with geoscientific by A. K Turner (ed.), NATO ASI Series, Kluwer Academic Publishings, pp. 215-232

Raper, J (1990b) Three-dimensional applications in geographic information systems. Taylor & Francis, London, 189 p

Raper, J (1993) Three dimensional GIS for the 1990. Seminar on Three Dimensional GIS - Recent Developments. ITC, Delft, The Netherlands, pp. 4-5

Raper, J (1989) The 3-dimensional geoscientific mapping and modelling system: a conceptual design. In: Raper, J (Ed.) Three dimensional applications in geographic information systems. Taylor & Francis, London

Raper, J (1990a) The 3-dimensional geoscientific mapping and modelling system: a conceptual design. In: Three Dimensional Applications in Geographic Information Systems, J. Raper (ed.) Taylor & Francis, pp. 11-19

Raper, J, Kelk, B (1991) Three-dimensional GIS, In: Geographical information systems: principles and applications. D J Maguire, M Goodchild and DW. Rhind (eds.) Longman Geoinformation, pp. 299-317

Requicha, AAG (1980) Representation for rigid solids: theory, methods, and systems, Computing Surveys. Vol. 12, No. 4

Rhind, DW (1992) Spatial data handling in the geosciences. In: Three-Dimensional Modeling with Geosciencetific Information Systems by A. K. Turner (ed.), NATO ASI Series C, Kluwer Academic Publishing, Dordrecht, Vol. 354, pp. 13-27

Richardson, DE (1993) Automated spatial and thematic generalization using a context transformation model. PhD Thesis. R&B Publications, Canada, 149 pp

Rikkers, R, Molenaar, M, Stuiver, J (1993). A query oriented implementation of 3D topologic datastructure. Proc. Fourth European Conference on Geographical Information Systems (EGIS'93). Genoa, Italy, pp. 1411-1420

Roessel, JW van (1986) Design of a spatial data structure using the relational normal forms. Proceedings of the 2nd International Symposium on Spatial Data Handling. Seattle, pp. 251-272

Rongxing Li (1994) Data structures and application issues in 3-D geographic information systems. Geomatica. Vol.48, No.3, pp. 209-224

Roushannejad, AA (1993) Mathematical morphology in automatically deriving skeleton lines from digitized contours. M.Sc. Thesis. ITC, Enschede, The Netherlands

Samet, H, Webber, RE (1988) Hierarchical data structures and algorithms for computer graphics: Part I – Fundamentals. IEEE Computer Graphics and Applications, May 1988, Vol. 8, pp. 48-68

Samet, H (1990) Applications of spatial data structures. Addison-Wesley, 507 p

Samet, H (1990) The design and analysis of spatial data structures. Reading, Addison-Wesley. Massachusetts

Sandgaard, J (1988) Integration of a GIS and a DTM. International Archives of Photogrammetry and Remote Sensing. XVI Congress, Commission III, Kyoto, Japan, pp. 716-725

Savarese DF (2003) Learning to Fly. Java Pro Magazine. June issue. http://www.fawcette.com/javapro/2003_06/magazine/features/dsavarese/.

Scott, MS (1994) The development of an optimal path algorithm in three-dimensional raster space. MSc Thesis. Department of Geograhy, University of South Carolina, 108 pp

Seed, GM (1996) An introduction to object-oriented programming in C++ with application in computer graphics. Springer-Verlag, London, 1048 p

Shamos, MI, Hoey, D (1975) Closest-point problems. Proc. of the 16th Annual Symposium on the Foundations of Computer Science (Washington: IEEE). pp. 151-162

Shekar, S, Chawla, S (2003) Spatial Databases – A Tour. Pearson Education. New Jersey

Shephard, MS, Schroeder, WJ (1990) A combined octree/Delaunay method for fully automatic 3-D mesh generation. International Journal for Numerical Methods in Engineering. 29, pp. 37-55

Shepherd, IDH (1991) Information integration and GIS. In: Maguire, DJ, Goodchild, MF, Rhind, DW (Eds.) Geographical information systems principles and applications - Vol. 1. Longman Scientific & Technical, New York, USA, pp. 337-360

Shibasaki, R, Shimizu, E, Nakamura, H (1990) Three dimensional (3D) digital map for an urban area. International Archives of Photogrammetry and Remote Sensing. Vol. 28, Part 4, Commission IV, Tsukuba, Japan, pp. 211-220.

Shibasaki, R, Shaobo, H (1992) A digital urban space model - a three dimensional modelling technique of urban space in a GIS environment. International Archives of Photogrammetry and Remote Sensing. Vol. XXIX, Part B4, Commission IV, Washington, D.C., USA., pp. 257-264

Shmutter, B, Doytsher, Y (1988) An algorithm to assemble polygons. ACSM-ASPRS Annual Convention. St. Louis, Missouri, pp. 98-105

Shi, W (1994) Modelling positional and thematic uncertainties in integration of remote sensing and geographic information systems. PhD. Thesis. International Institute for Aerospace Survey and Earth Sciences, Enschede, The Netherlands, 147 pp

Shi, W, Yang, B, Li, Q (2003) An object orientated data model for complex objects in three-dimensional geographical information systems. International Journal of Geographical Information System, Taylor & Francis, London

Sibson, R (1978) Locally equiangular triangulations. Computer Journal. 21, pp. 243-245

Sides, EJ (1992) Modelling three-dimensional geological discontinuities for mineral evaluation. PhD Thesis. University of London, 281 pp

Singer, IM, Thorpe, JA (1967) Lecture notes on elementary topology and geometry. Scott Foressman & Co., Illinois, USA, 214 p

Slingerland, R, Keen, TR (1990) A numerical study of storm driven circulation and 'event bed' genesis. Proc. of Symposium on Structures and Simulating Processes. Freiburger Geowissenschafliche Beitrage, 2, pp. 97-99

Sloan, SW (1987) A fast algorithm for constructing Delaunay triangulations in the plane. Advanced Engineering Software, 9, pp. 34-55

Smith, HC (1985) Data base design: Composing fully normalized tables from a rigorous dependency diagram. Communication of the ACM. Vol. 28, No. 8, pp. 826-838

Smith, TR, Menon, S, Star, JL, Estes, JE (1987) Requirements and principles for the implementation and construction of large-scale geographical information systems. International Journal of Geographical Information Systems. 1:13-32

Smith, DR, Paradis, AR (1989) Three-dimensional GIS for the earth sciences. In: Raper, JF (ed.) Three dimensional applications in geographical information systems. Taylor & Francis, London. pp. 149-155

Snyder, A (1993) The essence of objects: concepts and terms. IEEE Software. January, 1993, pp. 31-42

Sommerville, DMY (1929) An introduction to the geometry of N dimensions. Dover publications, Inc., New York. 196 pp

Special Interest Group (SIG) 3D (2004). Pilot 3D der GDI NRW – Ergebnisse. (URL:http://www.ikg.unionn.de/sig3d/docs/040109_Flyer_Endergebnis_3D-Pilot.pdf).

Stanat, DF, McAllister, DF (1977) Discrete mathematics in computer science. Prentice-Hall. Englewood Cliffs, NJ

Stoter J., and Oosterom P. van (2002) Incorporating 3D Geo-Objects into a 2D Geo-DBMS. Proceedings of ASPRS/ACSM. Washington, USA

Stoter, J, Zlatanova, S (2003) 3D GIS - where are we standing. Joint Workshop on Spatial, Temporal and Multi-Dimensional Data Modeling and Analysis. Quebec City, Canada

Stoter, J, Zlatanova, S (2003) Visualisation and editing of 3D objects organised in a DBMS. Proceedings of the EuroSDR Com V. Workshop on Visualisation and Rendering. Enschede, The Netherlands

Sun Microsystems (2004) The Java 3D API. http://java.sun.com/products/java-media/3D/.

Sutherland, IE, (1963) Sketchpad: A man-machine graphic communication system. TR-296, MIT Lincoln Laboratory, Lexington, Mass.

Sutherland, IE (1970) Computer displays. In: Beatty, JC, Booth, K, (Eds.) IEEE Computer Society Press. Silver Spring, MD, pp. 4-20

Taddei U. (2003) DEMViewer. http://www.geogr.uni-jena.de/~p6taug/demviewer/demv.html.

Takahashi, M, Yokokawa, T, (1992) The automatic selection system of transmission line routes based on DTM. International Archives of Photogrammetry and Remote Sensing. Vol. XXIX, Part B4, Commission IV, Washington, D.C., USA. pp. 883-885

Tang, L, (1992) Raster algorithms for surface modelling. International Archives of Photogrammetry and Remote Sensing. Vol. XXIX, Part B3, Commission III, Washington, D.C., USA., pp. 566-573

Tansel, AU, Clifford, J, Gadia, S, Seger, A, Snodgrass, R (1993) Temporal databases, theory, design, and implementation. Benjamin/Cummings Publishing Company, Inc., California

Tempfli, K (1986) Composit/progressive sampling - a program package for computer supported collection of DTM data. ACSM-ASPRS Annual Convention. Washington DC

Tempfli, K (1982) Notes on interpolation and filtering. ITC Lecture Note. 3rd edition, ITC, Enschede, The Netherlands

Tempfli, K, Makarovic, B (1978) Transfer functions of interpolation methods. ITC Journal 1978-1. pp. 50-78

Thiessen, AH (1911) Precipitation averages for large areas. Monthly Weather Review, July, 39, pp. 1082-1084

Thomas, D, (1989) What's in an object? *Byte.* March, 1989, pp. 231-240

TIGER, Topologically integrated geographic encoding and referencing system. U.S. Department of Commerce, Bureau of the Census

Tri-service data standard, (1994) CADD/GIS Technology Center

Tsai, VJD (1993) Delaunay triangulations in TIN creation: an overview and a linear-time algorithm. International Journal Geographical Information Systems. Vol. 7, No. 6, Taylor & Francis Ltd., pp 501-524

Tsai, VJD, Vonderohe, AP (1993) Delaunay tetrahedral data modelling for 3-D GIS applications. Proc. GIS/LIS '93 Conference, Minneapolis, Minnesota, Vol. 2, pp. 671-680

Tsai, VJD, Vonderohe, AP (1991) A generalized algorithm for the construction of Delaunay triangulations in Euclidean n-space. Proc. GIS/LIS '91 Annual Conference. Atlanta, GA, Vol. 2, pp. 562-571

Turner, AK (1989) The role of 3-D GIS in subsurface characterization for hydrogeological applications. In: Raper, JF (Ed.) Three Dimensional Applications in Geographic Information Systems. Taylor & Francis. London, pp. 115-127

Vaidyanathaswamy, R (1960) Set topology. Chelsea Publishing Company, New York

Voronoi, G (1908) Nouvelles applications des parameters continus á la théorie des formes quadratiques. Deuxiéme Mémoire: Recherches sur les parallelloedres primitifs. Journal fur die Reïne und Angewandte Mathematik. 134, pp. 198-287

Wang, ZJ (1994) Digital photogrammetric data acquisition for 3D GIS. M.Sc. Thesis. ITC, Enschede, The Netherlands, 88 pp

Watson, DF (1981) Computing the n-dimensional Delaunay tessellation with application to Voronoi polytopes. Computer Journal. 24, pp. 167-172

Watson, DF, Philip, GM (1984) SURVEY: Systematic triangulations. Computer Vision, Graphics and Image Processing. Vol. 26, pp. 217-223

Watt, A (1993) 3D computer graphics. Addison-Wesley Publishing Company Inc., UK, 500 pp

Webster, CJ (1990) The object-oriented paradigm in GIS. International Archives of Photogrammetry and Remote Sensing. Commission III, Vol. 28 Part 3/2, Wuhan, China, pp. 947-984

Webster, CJ, Omare, CN (1991) A formal approach to object-oriented spatial database design. Proc. of Second European Conference on Geographical Information Systems (EGIS'91). Vol. 2, Brussel, pp. 1210-1218

Weibel, R (1993) On the integration of digital terrain and surface modeling into geographic information systems. Proc. of 11th. International Symposium on Computer Assisted Cartography (AUTOCARTO 11). Minneapolis, pp. 257-266

Weiskamp, K, Flamig, B (1992) The Complete C++ Primer. 2nd. edition, Academic Press, Inc., USA, 540 pp

Willard, S (1970) *General topology*, Reading, Addison-Wesley, Massachusetts, USA. 369 pp

Wilson, RJ (1985) Introduction to graph theory. 3rd. Edition, Longman Scientific & Technical. UK

Wolberg, G (1990) Digital image warping. Los Alamo, IEEE Computer Society Press. Los Alamo. 318 pp

Worboys, MF, Hearnshaw, HM, Maguire, DJ (1990) Object-oriented data modelling for spatial databases. International Journal of Geographical Information Systems. Taylor & Francis Ltd., Vol. 4, No. 4, pp. 369-383

Vries, ME de, Zlatanova, S (2004) Interoperability on the Web: the case of 3D geo-data. IADIS International Conference on e-Society. Spain

Vries ME de, Stoter, J (2003) Accessing 3D geo-DBMS using Web technology.

Wachowicz, M, Bulens J, Rip, F (2002) GeoVR construction and use: The seven factors. Proceedings of the 5th AGILE. Palma

Web3D Consortium (2004) http://www.web3d.org/ (2004)

Worboys, MF, Hearnshaw, HM, Maguire, DJ (1990) Object-oriented modelling for spatial databases. Int. Journal of Geographic Information Systems (IJGIS). Vol. 4, No. 4, Taylor & Francis. London

Würländer, R (1988) Undersuchung zur Integration von digitalen geländemodellen in raumbezogene informationssystemme. Diplomarbeit. Technische Universität München

Youngman, C (1989) Spatial data structures for modeling subsurface features. In: Raper, JF (Ed.) Three Dimensional Applications in Geographic Information Systems. Taylor & Francis. pp. 129-136

Zeitouni, K, Cambray, B de (1995) Topological modelling for 3D GIS. 4th. International Conference on Computers in Urban Planning and Urban Management. Melbourne, Australia, http://www.prism.uvsq.fr/public/beatrix/publi_en.html

Zhu, C, Tan, EC, Chan, KY (2003) 3D Terrain visualization for Web GIS. Map Asia 2003, Kuala Lumpur, Malaysia, October 2003.

Zlatanova, S (2000) 3D GIS for Urban Development. PhD Thesis. ITC Dissertation Series No. 69 , The Netherlands

Zlatanova, S, Abdul-Rahman, A, Shi, W (2004) Topological models and frameworks for 3D spatial objects. Journal of Computers & Geosciences. May, Vol 30, Issue 4, pp. 419-428

Zlatanova, S, Abdul-Rahman A, Shi W (2002a) Topology for 3D spatial objects. International Symposium and Exhibition on Geoinformation (ISG). Kuala Lumpur, Malaysia

Zlatanova, S, Abdul-Rahman A, Pilouk, M (2002b) 3D GIS: Current Status and Perspectives. Proc. of the Joint Conference on Geo-spatial theory, Processing and Applications. 8-12 July, Ottawa, Canada

Index

Printing: Krips bv, Meppel
Binding: Stürtz, Würzburg